Creating an Effective Wireless Devi

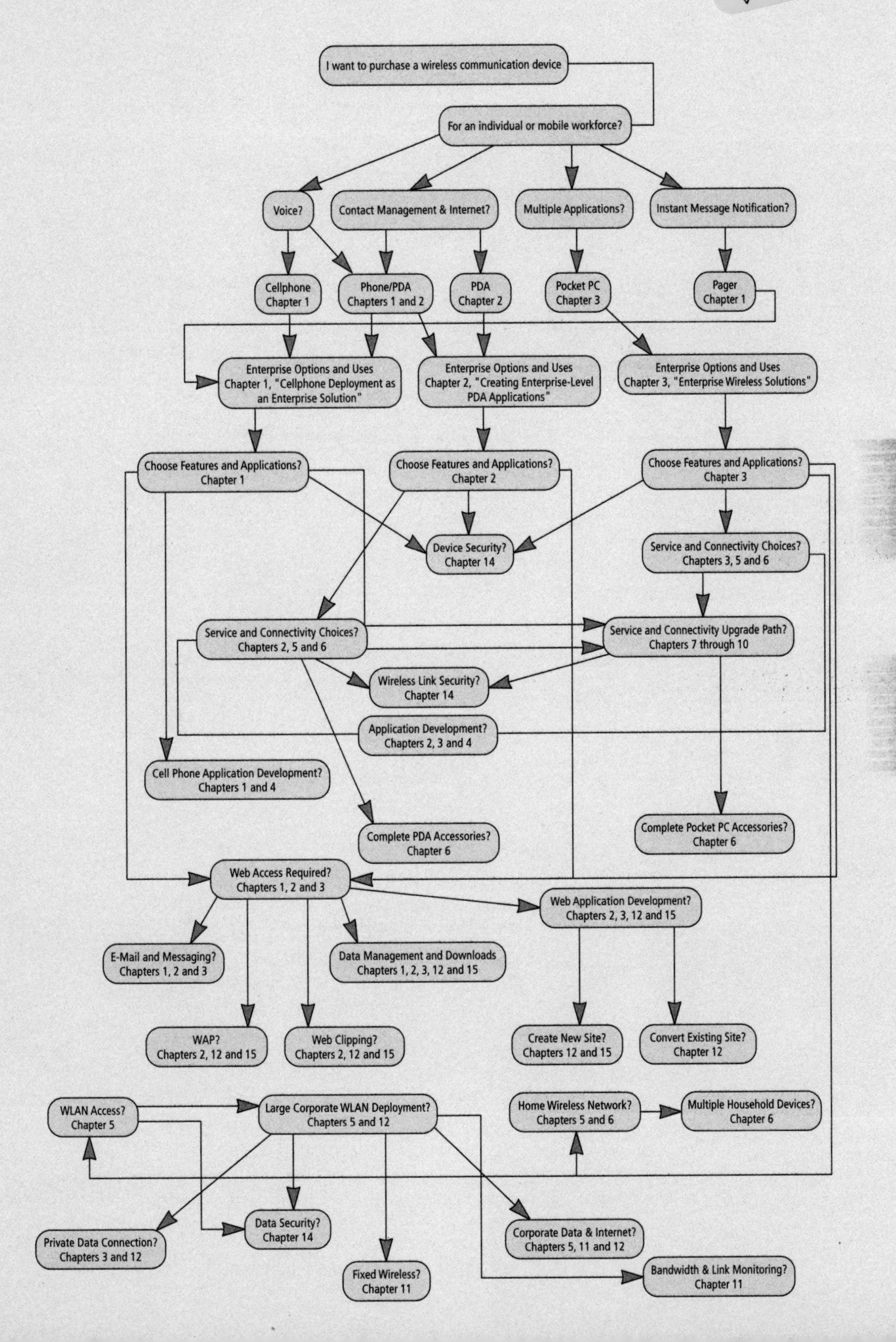

Wireless Devices End to End™

Wireless Devices End to End™

Winston Steward and Bill Mann

Ron Gilster, Series Editor

Best-Selling Books • Digital Downloads • e-Books • Answer Networks
e-Newsletters • Branded Web Sites • e-Learning

New York, NY • Cleveland, OH • Indianapolis, IN

Wireless Devices End to End™

Published by
Hungry Minds, Inc.
909 Third Avenue
New York, NY 10022
www.hungryminds.com

Library of Congress Control Number: 2001099740

ISBN: 0-7645-4895-6

Printed in the United States of America

10 9 8 7 6 5 4 3 2 1

1B/SQ/QS/QS/IN

Distributed in the United States by Hungry Minds, Inc.

Distributed by CDG Books Canada Inc. for Canada; by Transworld Publishers Limited in the United Kingdom; by IDG Norge Books for Norway; by IDG Sweden Books for Sweden; by IDG Books Australia Publishing Corporation Pty. Ltd. for Australia and New Zealand; by TransQuest Publishers Pte Ltd. for Singapore, Malaysia, Thailand, Indonesia, and Hong Kong; by Gotop Information Inc. for Taiwan; by ICG Muse, Inc. for Japan; by Intersoft for South Africa; by Eyrolles for France; by International Thomson Publishing for Germany, Austria, and Switzerland; by Distribuidora Cuspide for Argentina; by LR International for Brazil; by Galileo Libros for Chile; by Ediciones ZETA S.C.R. Ltda. for Peru; by WS Computer Publishing Corporation, Inc., for the Philippines; by Contemporanea de Ediciones for Venezuela; by Express Computer Distributors for the Caribbean and West Indies; by Micronesia Media Distributor, Inc. for Micronesia; by Chips Computadoras S.A. de C.V. for Mexico; by Editorial Norma de Panama S.A. for Panama; by American Bookshops for Finland.

For general information on Hungry Minds' products and services please contact our Customer Care department within the U.S. at 800-762-2974, outside the U.S. at 317-572-3993 or fax 317-572-4002.

For sales inquiries and reseller information, including discounts, premium and bulk quantity sales, and foreign-language translations, please contact our Customer Care department at 800-434-3422, fax 317-572-4002 or write to Hungry Minds, Inc., Attn: Customer Care Department, 10475 Crosspoint Boulevard, Indianapolis, IN 46256.

For information on licensing foreign or domestic rights, please contact our Sub-Rights Customer Care department at 212-884-5000.

For information on using Hungry Minds' products and services in the classroom or for ordering examination copies, please contact our Educational Sales department at 800-434-2086 or fax 317-572-4005.

For press review copies, author interviews, or other publicity information, please contact our Public Relations department at 317-572-3168 or fax 317-572-4168.

For authorization to photocopy items for corporate, personal, or educational use, please contact Copyright Clearance Center, 222 Rosewood Drive, Danvers, MA 01923, or fax 978-750-4470.

About the Authors

Winston Steward has written or co-written more than a dozen books on networking technology, computer graphics and Web design, consumer technology, and digital recording. Titles include *Scanner Solutions, Digital Video Solutions, Excel 2000 for Busy People,* and *CorelDRAW 8 Secrets.* He lives with his family in Los Angeles, where he develops software solutions for various non-profit organizations.

Bill Mann has been infatuated with computers since his high school installed a PDP-8 in the late 70s. He has published edutainment software, run his own software company, programmed military and commercial flight simulators, and written documentation for a wide range of software products. He has also written more than a dozen books and magazine articles on software, computers, the Internet, and the impact of computer technology on society. Based in Bedford, New Hampshire, Bill currently leads the documentation team at Pumatech, Inc., a leading provider of synchronization, page minding, and other personal information services for enterprises and the wireless community.

Ron Gilster, *End to End* Series Editor, is a best-selling author on the subjects of networking, PC hardware, and career certification. He has over 35 years of experience in computing, networking, and data communications, serving as a technician, consultant, trainer, and executive — most recently as the general manager of the Internet and backbone operations of a large regional wireless communications company.

Credits

ACQUISITIONS EDITOR
Katie Feltman

PROJECT EDITOR
Michael Kelly

COPY EDITORS
Gabrielle Chosney
Kevin Kent

TECHNICAL EDITOR
John A. Green

EDITORIAL MANAGER
Ami Frank Sullivan

SENIOR VICE PRESIDENT, TECHNICAL PUBLISHING
Richard Swadley

VICE PRESIDENT AND PUBLISHER
Mary Bednarek

PROJECT COORDINATOR
Maridee Ennis

GRAPHICS AND PRODUCTION SPECIALISTS
Melissa Auciello-Brogan
Sean Decker
Jackie Nicholas
Jacque Schneider

QUALITY CONTROL TECHNICIANS
David Faust
Carl Pierce
Angel Perez
Marianne Santy

BOOK DESIGNERS
Kathie Schutte and Michael Freeland

PROOFREADING AND INDEXING
TECHBOOKS Production Services

Preface

The hottest space in technology today is the wireless space. A few years ago, every company seemed to have a "dot com" of some sort in its name, or at the very least, an "Internet strategy"; today, however, every company seems to be creating something for the wireless space, or at the very least, creating a "wireless strategy." The reasons for this phenomenon are much the same as they were during the Internet frenzy.

Wireless devices, particularly digital wireless data devices (often called data terminals or mobile data terminals), can change the way you do business. Common examples of early adopters of wireless devices are sales forces in the field and senior executives who need to be in touch at all times. By using wireless devices with custom-inventory and sales-support applications, salespeople can check inventory, place an order, and close a deal — all without leaving the client's office. Executives equipped with always-connected e-mail devices such as the RIM BlackBerry can keep on top of rapidly changing situations while traveling or otherwise being outside the office.

Wireless devices are also infiltrating business in less well-known ways. People are beginning to shop wirelessly. Airlines are using wireless devices to confirm customer reservations from locations other than a ticket counter. Trucking companies and bicycle delivery services are using wireless devices to keep track of their rolling (or pedalling) stock. And companies are reducing costs and increasing their flexibility by connecting PCs to corporate networks wirelessly. Like the telephone, the PC, and corporate networks before them, wireless devices are becoming part of a newer, more efficient way to work.

Another powerful force driving wireless devices is the consumer market. Hundreds of millions of people around the world carry mobile phones. While the early market was built on analog phones with little or no data capability, the current generation of phones (called 2G and 2.5G) constitutes digital devices, and usually includes at least some data capability. In Japan, i-mode phones and the tens of thousands of i-mode sites with which they function are wildly popular, particularly among young people. In Europe (and increasingly Asia and North America), GSM (Global System for Mobile Communications) phone users send billions of short text messages to each other every month.

To give you some idea of the wireless device growth projections that are being published in the summer/fall of 2001, note the following:

- Japanese cellphone conglomerate NTT DoCoMo predicts that by 2010 in Japan alone there will be up to half a billion mobile data terminals (cellular phones with data capability and similar devices) in use, including up to 40 million refrigerators and 20 million devices worn by domestic pets.
- The International Telecommunications Union (ITU) predicts one billion cellular telephone subscribers by 2002. The Strategis Group expects 172 million mobile data device subscribers in the United States alone by 2007.
- NTT DoCoMo reported almost 27 million i-Mode users as of August 2001. (Keep in mind that the population of Japan is around 130 million.)
- The GSM Association predicts that over 200 billion SMS (Short Message Service) messages will be sent in 2001.

- The wireless research organization EMC World Cellular Database reported that in May of 2001, 1 in 12 human beings carried a GSM phone.

Why We Wrote This Book

A few years ago, the publishing world was rather infatuated with the phrase, "on Internet time." Taken from the title of a book chronicling the Netscape/Internet Explorer browser wars, the phrase has been used to describe the breakneck pace of product development and rising consumer expectations in the Internet age. However, "Wireless time" occurs in cycles that make Internet time seem positively Jurassic. The development cycles described during that period (the late 1990s, or "ancient history") and the regularity with which consumers were asked to pony up for a new round of purchases are nothing compared to the speed with which the wireless phenomenon has overtaken us. Technology is released and products are developed, and we must all respond quickly: Yes? No? Maybe? Yes, but wait?

And at each decision point, much is at stake. A timely wireless technology purchase could save a company millions of dollars, give it a priceless edge over the competition, or leave that company saddled with obsolete equipment before the year's end. For consumers, a wisely considered wireless purchase could streamline life and reduce drudgery in ways that, even six months ago, were not at all possible. However, a poor purchase could leave you foundering with a subscription to a discontinued service, or one that runs at a snail's pace compared to your ultra-hip coworkers. Not exactly wreckage and ruin, but no one wants to spend $300 on a doorstop.

This book, then, is your guide to wireless devices and their driving technologies, which, taken together, have created the fastest-changing consumer market of all time. This book is your compass to help you navigate information about all your wireless purchases, great and small. We provide guidelines to help predict which wireless technologies are likely to be viable years from now, as well as how to purchase a cellphone, PDA, or Pocket PC with a bit of confidence. We don't tell you what to buy or who tomorrow's big winners will be. Rather, we level the playing field on your behalf. You learn how to spot trends, develop reasonable upgrade paths for your wireless purchases, and avoid being caught off guard by tomorrow's emerging wireless technologies.

Who Should Read This Book

This book is written for two types of technically oriented readers. The primary audience is the technical decision-maker in any size corporation — the person who must define the company's wireless strategy. If you are such a technical decision-maker, you will benefit from the information in all three parts of this book. You need to know what kinds of wireless devices are out there now and have a basic understanding of the technologies those devices are built on, as well as some idea of where wireless devices are headed.

The secondary audience is a technologically sophisticated mobile professional who wishes to understand and take advantage of the latest wireless devices. If you are such an individual, you will likely find Parts I and III the most useful. You will probably want to know what wireless device to buy (or buy next), and what kinds of cool technology are on the horizon.

What You Need to Know to Read This Book

There are no prerequisites for reading this book, although a general understanding of the Internet and PCs, as well as a general comfort level with new technologies, will certainly help.

If you are not at all technically inclined, you will probably want to avoid Part II. While we've tried to make the technological discussions as easy-to-read as possible, you may still find some of those chapters heavy-going.

What the Icons Mean

This book is studded with various icons in the left margins of the pages. These icons denote various types of information that warrant special attention as you read the book. The icons include the following:

The Caution icon indicates potential problems and ways to work around them, if they exist.

This icon indicates that material relevant to what you are reading is available in another section of the book.

This icon indicates where material relevant to what you are reading is available on the Web.

The Tip icon indicates ways to work more efficently, or a new approach to performing a task.

The Note icon provides additional information that may be useful to you as you make your wireless decisions.

How This Book Is Organized

The book is divided into three parts. The first part discusses real wireless devices that you can use today. The second part provides information about many of the important technologies that constitute the devices in the first part. The third describes various applications, issues, and wireless concerns not covered in the first two.

Part I: Wireless Devices

Part I provides an overview of the kinds of wireless devices that are available today.

Chapter 1 covers cellphone types, brands, and technologies, providing insight into what purchase may be best for your needs. We discuss criteria for making an informed cellphone purchase or mobile deployment decision. Pagers are also discussed.

Chapter 2 discusses personal digital assistants, again focusing on product types, technologies, and criteria for choosing a PDA for individual use, as well as deployment of many units to a mobile workforce. Special emphasis is placed on enabling your PDA as a wireless communication device. Case studies are presented describing the benefits of a PDA-enabled mobile work team. The Palm OS, software selection, and supplemental products are also discussed.

Chapter 3 discusses Pocket PCs and the Microsoft Pocket PC OS, again with an emphasis on selecting the right unit, enabling it for wireless communication, selecting software, and issues for multiple unit deployment for business. We discuss examples of how Pocket PCs have been incorporated into many working environments.

Chapter 4 describes the Symbian OS and Java-enabled cellphone technology, which have been embraced by many as holding great promise in device-level international wireless communication. You'll learn why these flexible and scaleable technologies play a large role in personal wireless devices worldwide.

Chapter 5 describes Wireless Local Access Networks (WLAN), their underlying technology, how they are deployed, as well as large- and small-scale technology and equipment considerations.

Chapter 6 rounds up a broad collection of wireless products and considers how they may fit into your personal and corporate life.

Part II: Techs and Specs

Part II provides a more in-depth look at many of the key technologies upon which the devices in Part I are built. If you are the technologically sophisticated mobile professional who wants to select a new wireless device, this part of the book will probably interest you less than it would the technical decision-maker, who will be building wireless into the corporate strategy.

Chapters 7 through 10 cover the *generations* of wireless technology. These generations, denoted by the terms 1G, 2G, 2.5G, and 3G, are shorthand means of identifying four levels of performance and capability in wireless wide area networks.

Chapter 11 moves away from mobile wireless devices and introduces a discussion on fixed wireless technology. Various methods for bringing wireless communication to differing population settings will be discussed. You'll read about point-to-point, point-to-multipoint, last mile, and other fixed wireless technology issues.

Part III: Sundry Technologies, Products, and Developments

Part III of the book is dedicated to the future of wireless technology. In this section, visionaries, researchers, industry observers, and others who have influence over the future of wireless devices have provided their thoughts on what our wireless future will look like. If you're as interested in where we are going as in where we are today, you should find this part of the book quite interesting.

Chapter 12 poses a number of common wireless technology questions and concerns and provides a variety of possible answers. Readers are directed elsewhere in the book for more detailed answers and discussions.

Chapter 13 looks at the technologies behind the diverse devices covered in Chapter 6. This chapter also discusses some of the technologies that will find their way into future wireless devices, including more readable displays and more potent power sources.

Chapter 14 addresses a subject that is on the minds of many people in the wireless industry today, namely wireless security. The chapter surveys the kinds of security issues surrounding wireless devices and describes some steps you can take to improve security on a wireless LAN.

Chapter 15 discusses the wireless Web and describes the special needs and requirements for creating wireless-friendly Web sites. You'll learn about wireless Web technology and the various ways to develop a site easily accessed by PDAs, Pocket PCs, and data-enabled cellphones.

How to Use This Book

This book is designed so that you can read individual chapters in any order, with Cross-Reference and On the Web icons referring you to additional information elsewhere in the book or online. Here are some additional ways to read the book, based on your individual goals:

- If your main interest is learning about the wireless devices that are available today, you may want to jump into Part I and start reading.
- If you are involved in setting your organization's policies on wireless devices, you should probably read the book cover to cover, to get a comprehensive overview of the field.
- If you have general questions about wireless devices, look in the Appendix for questions that most closely match yours. With a little luck, it will point you to the right part of the book. If nothing else, you should be able to find terms related to your area of interest that you can then look up in the Index and Table of Contents.

Acknowledgments

Winston Steward would like to thank Katie Feltman and Ron Gilster for the opportunity to be part of this project. This is truly a book that needed to be written. Special thanks to my agent, Margot Maley, my family, and my best friend, Louna.

Bill Mann would like to thank my agent Margot, who got me exactly the right project at exactly the right time; the rest of the team at Waterside for all those things they do to make my book projects happen; Ron, for his overall role as series editor and for coming up with a book concept that involved my playing with so many wild wireless widgets; the team at Hungry Minds, including Mike Kelly, Gabrielle Chosney, and Kevin Kent, for the massive behind-the-scenes effort involved in turning a whole lot of words and images into a real printed book; and everyone who sent me wireless devices to work (and play) with — you made this a much better book (and made this gadget freak quite happy). Finally, I want to thank Patti and Jenn — Jenn, for giving up a lot of our time together once again and for being such a great kid through it all; and Patti, for her patience and support through this project, not to mention the dozens of quality line drawings she rendered out of my scribbled notes. You guys were great!

Contents at a Glance

Contents

Introduction

What Is Wireless?

Wireless technology is the ability to transmit a message via a carrier wave. The sent information is combined with the wave, or modulated. Then, as the information is received, it is demodulated and extracted for use. The information can be voice or data, analog or digital; it can be a complex message or a simple on/off command. Wireless has been around since the dawning of the electronic age, when Guglielmo Marconi first successfully transmitted information over radio waves. Since that time, its basic aim has changed little: to find available radio spectrum and send a message from here to there.

That said, the term "wireless" encompasses a plethora of technologies — some new and some old. A few have just been implemented; nonetheless, they hold incredible promise.

The Wireless Future and You

This book describes prevalent wireless devices — their applications and platforms, as well as how they are changing. The prospect of finding more efficient methods with which to send varied types of information has again made the world excited about technology industries. Commentators are quick to predict a wireless world, offering utopian convenience and agility to every major industry and household. However, it's clear that such advances will come in bits and pieces.

Some of these advances will be of interest to you, some won't. As this book takes you through the ins and outs of wireless devices, it will help you find your niche. What should you buy? What should you deploy? What should you bet or bank on? Whether you want to discover the best personal mobile communications device for your needs or oversee the deployment of a wireless platform for your enterprise, this book will simplify the array of choices.

Moving into the "New" Wireless

In part, the dizzying rate of change in wireless industries was the impetus for the writing of this book — to document and explain the latest improvements. Yet some aspects of wireless technology don't need to change. Much under- and aboveground equipment does not need to be replaced, and some device types, such as CB radios, Private Mobile Radio systems, and Family Radio Service systems, are working just fine, thank you, and have been for decades. So in this introduction discusses the drive behind the big changes and why they happen so fast, and yet somehow, not fast enough.

Those aspects of wireless that experience rapid change are those that carry the weight of expectation. We've all heard the promises: traffic signals that change to accommodate emergency vehicles, alarm clocks that ring a few minutes early if the Web reports heavy local traffic, or a mortgage broker's ability to update a real estate listing and have every agent's handheld device updated as well.

However, each new innovation requires several leaps in technology, and each new advance summons additional obstacles. Indeed, the history of wireless is a history of overcoming hurdles. Before a charger cradled your cellphone, many roadblocks (such as overcoming data speed limitations) had to be overcome. To empower that cellphone to notify you of your kids' whereabouts, or to audibly notify you when the Atlanta Braves win a game, a few more obstacles will need to be overcome. But believe me, all major industry players are working on it.

Wireless technology is deployed unevenly. Japanese teenagers have been surfing the Web from their i-Mode phones for quite some time now. Europeans carry their cellphone identities on a *Subscriber Identity Module* (*SIM*), plopping it into a borrowed or rented phone as they travel from country to country. And the *Short Messaging System* (*SMS*), which enjoys much popularity among cellphone users worldwide, has not yet caught on in the United States. In this book, we talk about these and other impediments to wireless advances. As we discuss a wireless product, platform, or technology, you'll gain insight as to whether you should jump into the market now, jump later, or simply save your money and wait for the next big thing.

Riding the Waves Efficiently

To gain insight into what's happening now, we look at how one early wireless hurdle was overcome: achieving interference-free radio transmission.

As you know, there are two fundamental broadcast radio technologies: AM and FM. *AM,* or *amplitude modulation,* carries signal by varying the amplitude, or height, of the carrier wave. However, interference can easily cause random changes in an AM signal, resulting in poor transmission.

Thus, *FM,* or *frequency modulation,* was developed. FM carries information along a wave by varying the speed, or frequency, of the carrier wave. This broadcast method is less susceptible to interference, making FM preferable for broadcasting music.

However, one basic rule of radio transmission is that lower frequencies travel longer distances. Thus, AM, which broadcasts on relatively lower frequencies, is still popular for long-distance radio, including short-wave radio, which employs receivers that can pick up radio broadcasts from thousands of miles away on very low frequencies. This is also why you seem to pick up AM stations before FM as you approach the outskirts of a town.

One spin on FM transmission is to not "use up" an entire wave to send information, but to suddenly shift the position, or phase, at which the wave is transmitted, enabling you to send several messages in a single wave. This type of transmission is called phase modulation (PM). PM transmission is not well suited for analog broadcast, such as voice, music, or any transmission type that requires continuous waves. However, PM is great for digital wireless transmission.

Unforeseen Developments

Sometimes, the effect of implementing a particular technology is not evident till many years have passed. When the FCC first granted licenses to cellphone companies, the companies were given permission to broadcast in the 800 MHz range. This is a relatively low frequency, as frequencies suitable for voice communication go. Thus, providers could place cellular base stations fairly far apart, taking advantage of the longer distances that lower frequencies could transmit. As more companies applied for licenses, newer companies were granted permission to broadcast in the 1900 MHz range. Since higher frequencies don't transmit as far, providers were required to build base stations closer together, an initial financial hardship. However, higher frequencies can accommodate more calls; so as airwaves got crowded, companies that had higher frequency lines could accommodate more customers, and the initial investment paid off.

Roadblocks and Advances

For the remainder of this introduction, we discuss the technical advances that apply primarily to cellular phones, PDAs, handhelds, and mobile computers. As you know, the distinction between these devices is diminishing, and their requirements all drive the wireless industry to the pursuit of *spectral efficiency*. Spectral efficiency is the goal of allowing more and more customers to simultaneously use the same segment of bandwidth. Spectral efficiency enables more calls to go through, and more data to pass through the system faster. This requirement works hand in hand with another major goal of wireless—that of allowing customers to share both voice and data over the airwaves. How is this implemented? Well, the need to accommodate voice and data traffic at ever-increasing speeds dovetails nicely with the process of moving mobile wireless communication from the analog to the digital domain. These are the big advances of our time—the stories that will make tomorrow happen.

Multiplexing

Spectral efficiency is a huge focus of telecommunications—and the wireless industry—today. It is usually talked about as some form of multiplexing, or data-transmission methods that allow as many people as possible to use the same segment of bandwidth.

In analog telecommunications, spectral efficiency can be achieved by subdividing the spectrum into as narrow a band as possible, maximizing the number of callers that can access it.

In digital, two methods of multiplexing exist: dividing the spectrum into timed slots, and encoding each transmission so that more than one caller can use the same airwaves simultaneously. The two broad approaches to digital transmission of telecommunications data are of great interest to us today. The first is *Time Division Multiple Access* (*TDMA*), which is currently implemented on phone systems all over the world. It divides a spectral band into several time slots and calls transmit data at each instance of its own assigned time slot. Its implementation is what allows portions of the American-based AMPS phone system to be more or less "digital."

The second approach is called *Code Division Multiple Access* (*CDMA*). Using CDMA, all calls transmit data on a spectral band simultaneously, but each data segment is encoded, transmitted, and decoded by its own intelligent receiver. CDMA, which is gradually being implemented, has been enthusiastically received as the best technology for transmitting voice and data together. However, initial implementation costs will be high, since TDMA technology is already in place. Many have argued that upgrading TDMA technology is more cost-efficient than implementing costly CDMA systems.

Digital Communication Systems

Digital wireless communication has the following advantages:

- ✓ Digital data is easier to encrypt, making it more secure.
- ✓ Digital data is easier to compress, which means more data can be transmitted.
- ✓ Digital data makes it easier to correct transmission errors.
- ✓ Digital data makes it easier to combine voice and data transmissions.
- ✓ Digital data can be sent in packets (small bits of information sent in bursts); analog data is sent along a continuum. Data sent in packets can arrive at its destination out of sequence with other parts of the same message. This freedom of data flow increases efficiency and speeds transmission.

Dual- or Single-Channel

The way bandwidth is licensed and utilized with regard to digital data transmission — rather than just analog voice — is being reconsidered. Analog voice transmission requires duplexing — in other words, bandwidth is set aside for both talking and listening. To accommodate the needs of voice transmission in the United States, the FCC licenses bandwidth in paired frequencies. The frequency that the mobile device transmits is called the *uplink*. The frequency that the base station transmits is called the *downlink*.

FDD versus TDD

The most common and widely implemented technology for cellular voice transmission is *FDD,* or *Frequency Division Duplex*. FDD uses two separate frequency bands and also sets aside a guard band between these two bands to prevent cross talk between the uplink and downlink bands. However, digital enthusiasts point out that this is an awful lot of spectrum to set aside for a simple phone call.

Thus, for transmitting digital data (which does not require channel duplexing), a system like FDD is wasteful. *TDD,* or *Time Division Duplex,* is more efficient because it uses one channel for transmitting and receiving, and no guard band. This allots less spectrum to a single communication instance.

Currently, TDD is used primarily in a LAN environment for peer-to-peer communication. Nonetheless, because of its promise, advocates want it deployed more broadly as a way to save spectrum. Again, one industry-wide discussion is whether it is more costly to upgrade FDD systems to increase their efficiency, or to switch to TDD technology. Investing in TDD may make more sense, as more data transmitted via the airwaves is digital.

Synchronous versus Asynchronous

Telecommunications futurists point out that users will eventually expect to surf the Web and send e-mail on the same portable device on which they talk. They also point out that while phone conversations may require equal amounts of two-way bandwidth, Web surfing does not. After all, when you view Web pages or read e-mail, you are downloading images and documents, while "uploading" only a few mouse-clicks. Web communication, then, is *asynchronous,* and efficiency demands that systems do not "lock in" two open circuits for a Web-surfing session. Moving towards asynchronous data transfer would be spectral-efficient, but would also require a change in the way voice is transmitted via airwaves.

Circuit Switching versus Packet Switching

Wireless industries seek ways to transmit voice data similar to the way data moves through the Internet. The transition, however, is not so simple. By nature, voice data is analog. Voices are waves, and like all sounds, they flow along a continuum. Analog phone systems transmit data as a continuum as well. Hence, on analog phone systems, the beginning of a message cannot be separated from its ending. During a phone conversation, circuits must remain continuously open to accommodate the flow of conversation. Until very recently, telephone systems of all types operated using *circuit-switching* technology. A phone call is connected, and circuits are switched on. The circuit (and there are physical circuits involved in this process) stays continuously connected until the call disconnects.

On the other hand, data on the Internet is sent in packets. An e-mail or image will be broken down into thousands or millions of packets, each encoded with its destination (which means it doesn't need to arrive in any particular sequence with regard to other packets that are part of the same message). The Internet is a communication medium that does not prioritize one destination over another. The packet of information that arrives at a transmission point first will get sent first, regardless of destination. This process is efficient in terms of bandwidth usage, but there is no guarantee that the millions of packets of data that make up your simple e-mail message will not get there all at once. This method of data transmission, called *packet switching,* is desirable for its efficiency; packet switching allows a circuit to minimize a connection to the smallest amount of bandwidth until packets of information arrive and require more bandwidth.

HSCSD versus GPRS

The wireless industries are gradually implementing technology that can transmit voice traffic in a packet-switching environment. This achievement would lead us much closer to the telecommunications Holy Grail of wireless communications, where customers could send e-mail and data simultaneously with voice conversation.

One new technology that greatly increases transmission speed and facilitates voice and data transmission on currently existing phone systems is *High Speed Circuit Switched Data,* or *HSCSD.* HSCSD allows the use of four simultaneous channels. Data can be asymmetrical, so it is somewhat efficient for wireless Web surfing. Again, HSCSD is a technology that is available now, but it has its share of problems. Because it is based on circuit switching, it has some built-in bandwidth waste, based on the number of circuits it must keep open in order to facilitate its high-speed transfer performance.

The better, up-and-coming alternative to HSCSD is *General Packet Radio Service,* or *GPRS.* It is a true packet-switching system that provides "voice over data." GPRS can keep a device connected using very little bandwidth, so it facilitates the very desirable "always on" condition, giving mobile device users instant Internet access whenever their device is with them. Always on also facilitates instant notification of e-mail and other events. You'll be hearing more and more about GPRS and its variations in the years to come.

Part I

Wireless Devices

Chapter 1

Cellphones and Pagers

In This Chapter

The following topics are covered in this chapter:

- ✓ How cellphones work
- ✓ Features common to all cellphones
- ✓ Select cellphone features
- ✓ Accessing the Web via cellphone
- ✓ Cellphones and e-mail
- ✓ Considerations before signing up for a cellular service provider
- ✓ Cellphones as a business solution
- ✓ Pagers and messaging devices
- ✓ Two-way pagers
- ✓ Wireless e-mail devices

The cellphone, which brings together a dazzling array of technologies in a single device, has become inseparable from modern life. For many, cellphones are a lifeline, establishing and affirming a constant connection to anyone, anywhere.

Some of the technologies and electronics used in cellphones are familiar to many — for example, the LCD that displays alphanumeric characters, the antenna that receives radio frequency signals, and the amplifier that strengthens those signals.

Other technologies found in cellphones are the target of constant improvements. Microprocessors and analog-to-digital converters must be made ever smaller and more efficient, and the Digital Signal Processor (DSP) that compresses and decompresses voice data must recreate voices ever more accurately. Additionally, the technologies that carry the data from point-of-origin to destination must be configured to move more types of data at faster and faster rates.

Initiating a Cellphone Call

A cellphone may properly be called a terminal, because it is activated as the end result of a long chain of signals sent across a network. When you turn on your cellphone, it scans and then tunes in to an available communication channel, relaying its system I.D. number (SID) to the network. The network indicates which channel to use for receiving calls. When a cellphone receives a call, control data is shared between network and cellphone, establishing identity, access information, and channel assignment. The phone stays connected by constantly relaying its location to the network and by always tuning in to the nearest "cell" (transmission tower).

Routing a Call and Maintaining a Connection

Calls are routed through a base station and switching centers. These switching centers separate voice information and control data from its carrier wave, as well as allow cellphones to receive calls from the public phone system (PSTN). The closer the phone is to a transmitting cell, the stronger the signal. Therefore, throughout the course of the call, the cellphone is constantly relaying its location, maintaining the call by always tuning in to the nearest transmission tower (cell) in the network.

More Data — Faster

Of course, newer cellphones transmit more than voice data. Text messages, Web pages, and other types of digital information, such as notification of selected events, are all relayed via the network's cells. The complexity and variety of this information requires an ever-growing commitment on the network provider's part to build new transmitters capable of sending and converting the information into a form that the cellphone can use.

In short, data shared between cellphones and networks must be compressed, converted, routed, and relayed in many ways before it reaches a destination. In later chapters, you'll learn more about the systems and technologies that make this possible. For example, in Part II, you'll read about the migration from America's original analog AMPS (Advanced Mobile Phone System) to the digital and more global GSM (Global System for Mobile Communications). For now, the discussion in this chapter focuses on different cellphone features, how they work and interact, and how to pick the right phone for you.

Features Common to All Cellphones

This section describes the basic features found in most, if not all, cellphones. In addition to having an earpiece, speaker, and call indicator, all cellphones allow you to store names and numbers in *memory slots*. Some phones let you store fewer than 100 numbers, while others provide for hundreds. With some advanced phones, you can store multiple numbers for one name, as well as associate an e-mail address with a name. Very advanced phones (such as the Ericsson R380, Sprint Kyocera QCP-6035, and Samsung SPH-I300) blur the line between cellphone and PDA, allowing you to save complete contact information with a phone number.

Call Logs

All cellphones keep call logs, allowing you to view lists of missed calls, received calls, and call lengths. You can also reset call timers at the start of your billing cycle to keep track of phone use you'll be charged for.

Multiple Ring Tones

These days, all cellphones provide multiple ring tones. You'll be able to scroll through a list of tone combinations and classic melodies and select one as your ringer. Most phones let you download ring tones, such as popular song snippets and sound effects. Some phones provide silent call notification. Others, such as the Nokia 3210, let you compose ring tones, creating your own personalized ring using an electronic music system called MIDI (`http://ringtones.mobile-phone-guide.co.uk/work/compose.htm`). If you want a small orchestra to alert you when you are called, you'll enjoy this feature. And finally, some phones, such as the Ericsson R289, provide differential rings, letting you assign rings to specified callers.

Menu Navigation Keys

All phones have some sort of navigational menu for displaying sets of options — for example, Phone Book, Messages, Call Log, Settings, and so on. An easily navigable menu can make or break your satisfaction with your cellphone. Saving a new contact, or accessing a frequently called number, should all happen in a few keystrokes.

Basic Phone Features

All cellphones provide basic phone services such as call-waiting notification, automatic redialing, last number dialed, and voice and ringer volume. Additionally, any phone will let you put a caller on hold while you swap to an incoming call and will display a caller's number (unless they've placed a block), allowing you to send them to voice mail, if you wish.

Oddly though, not all cellphones come with a mute button or speakerphone. The Samsung SCH-3500 and the Sanyo SCP-4000 are two examples of phones that do.

Visual Reception and Battery Indicators

All phones provide a visual signal indicator so that you'll know not to make a call from a location with a very weak or nonexistent signal. You can also count on a clear indicator that you've entered an out-of-network area and are about to be charged for *roaming*. Also standard are battery strength indicators and low-battery beeps.

Phone Charger

You won't have to pay for a single phone charger, although a charger for your car will probably be extra. And with all the increased concern regarding cellphone use and auto safety, count on a hands-free headset being thrown in as well.

Adequate Reception and Call Clarity

Regarding phone service, you have the right to expect adequate call clarity and decent reception in those regions in which your provider purports to offer service. However, everyone has bad days. Calls will be dropped, and no doubt you'll have more problems during peak hours. (You always wondered why they gave you those free nights and weekend minutes, right?) Nonetheless, using your cellphone should not be a constant struggle to maintain a connection.

Basic Voice Mail

Voice mail — a system that allows callers to leave you messages 24/7, even when your phone is off — is a standard service provided with all cellphones. The number of messages your voice mail will hold varies from service to service. Some phones, such as the Nextel i85 series, require you to sign up with specific service providers when the phone is purchased. This may be advisable when buying a particularly high-tech phone, since many advanced services, such as Web browsing and e-mail notification, require specific service technologies.

Select Cellphone Features

This section describes select cellphone features, those that are not available in all phones. You may want to identify the select features that you would like to have and then do a little shopping around. Perhaps your dream phone is out there.

Some of the features listed in this section will be available in most phones, and you'd be hard-pressed to find newer phones that do not include them in some form. Others are quite specialized and have yet to catch on in any big way. Of course, as trends change, today's "extra" becomes tomorrow's basic standard.

Text Messaging

Most phones provide some sort of *SMS*, or *Short Message System*, capability. Sending text messages via cellphone is gaining in popularity. Text messages are discreet and timesaving. There's no reason why your basic "Dinner at 6:00" message should require the ritual of a phone call.

The number of lines of text and characters allowed per message varies greatly from phone to phone. The Ericsson R290(see Figure 1-1) is a particularly text-friendly phone, including a chat feature and the ability to plug an optional keyboard directly into the phone for easy messaging.

Before buying an SMS-enabled phone, make sure that the phone can both send and receive text messages. A one-way text message feature is of limited value.

Following are other typed message features:

- ✓ Some phones provide adjustable fonts, allowing you to compose messages with more lines of text (as long as your eyes can read the small lettering).
- ✓ Many phones feature *predictive text,* anticipating an intended word based on its opening letters. Keep in mind that with few exceptions, you'll be typing text on a tiny phone keyboard, so you'll appreciate any help completing common words and phrases. You'll find that, as part of the trend towards providing room for working with text, cellphone displays are getting larger as well.

- ✓ Some phones include preset messages that are accessible by scrolling, for example, "Confirming our plans," or "Urgent. Please call at once."
- ✓ On some advanced phones, the message recipient receives instant notification and display of the message.

Figure 1-1: The Ericsson R290 (photo courtesy of Telefonaktiebolaget L M Ericsson)

Color Display

Some cellphones enable you to download images and surf the Web, two experiences that are much more rewarding if done in color. Therefore, you should see more and more phones providing color displays. The color depth will not be anything near what a computer monitor, or even a handheld PC, displays. As of this writing, few color-display phones have been released to the U.S. market, with one exception being the Ericsson T68.

Short Text Messages Are Not E-mail

As more and more manufacturers produce devices that combine mobile phone and Internet-enabled PDA features, it's important to clarify the difference between sending and receiving text messages and working with e-mail. The Short Message System that enables you to create simple text messages is embedded in the digital technology used to send and receive voice calls. In other words, it's just another layer of data that lies on top of the phone connection. E-mail is a completely different system, involving the application of an additional layer of network protocol. In other words, it's not part of your phone call. Although most cellphones let you send text messages, only a growing number of cellphones let you send e-mail. E-mail and short text messages are not the same thing.

Dual- or Multi-Mode

Some phones let you choose which channel to use, a feature that is applicable only in areas where there is a choice. For example, in some locations using GSM, you can choose between 800 and 1900 MHz bandwidth, or you can switch between analog and digital, or between a local digital bandwidth and the worldwide GSM. Again, this feature depends on location as well as phone capabilities. World phones such as the Sprint Kyocera, the Motorola V60c, or StarTAC, and the Panasonic Versio provide mode switching.

Hands-Free and Wireless

If you desire a cellphone with a wireless headset, check out the Ericsson T68 phone, the Nokia Bluetooth Headset (see Figure 1-2), or the Plantronics M series wireless headsets for cellphones. The Ericsson phone's wireless headset uses Bluetooth technology, so it can connect with a host of other local wireless devices, using the same wireless hub.

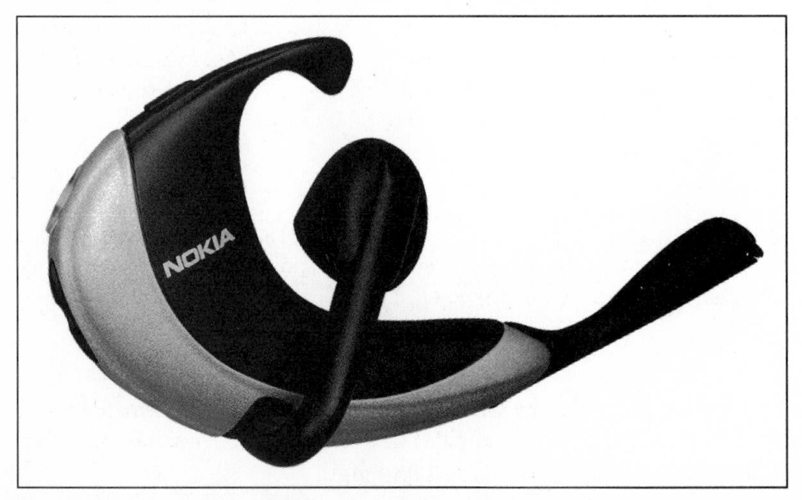

Figure 1-2: The Nokia Bluetooth Headset (photo courtesy of Nokia)

Backlit Keys

Backlit keys may not seem like a big deal until you have to dial a number in the dark. If you think that you'll be using your cellphone frequently at night, or in minimal light, try to get a phone with backlit keys. The Sprint Touchpoint 2100 is such a phone.

Voice-Activated Dialing

This feature, available on Samsung and other Sprint phones, can be quite a treat. First, you program the phone with a number, and then you speak the person's name attached to that number. You may have to say the name a couple times, in a clear voice, before the phone will recognize it, but after recording, you can dial the number by simply speaking the person's name.

Programmable and Differential Ring Tones

Programmable ring tones may seem like a luxury, but if you've ever been in a restaurant, heard a cellphone ring, and seen everyone reach simultaneously for their pocket or purse, you can imagine why unique rings would be helpful.

Differential rings let you know who is calling without even viewing the number.

Cellphones as Music Devices

Some cellphones include MP3 players or radios, allowing you to listen to music via a headset. You can store only a handful of MP3s on your phone, and, of course radio reception depends greatly on location; nonetheless, for the mobile music enthusiast, this feature provides two devices in one. Look into the Samsung Uproar M100 (see Figure 1-3), the Samsung SPH-M100, or the Ericsson R289LX.

PDA Options

A handful of phones (this number is growing), provide PDA options such as detailed and sortable contact information, a scheduling calendar and organizer, a to-do list, a calculator, and even handwriting-to-text conversion. These features come at a price, of course, and to be useful, the phone would require a screen much larger than a typical cellphone LCD. A large screen, however, pushes the device upward in size and weight, making it much bulkier than most would prefer for a cellphone. The Ericsson R380 mitigates some of this by providing a keypad that flips down to reveal a PDA-sized display, which is used by turning the phone on its side.

Games

As phones have grown more sophisticated, so have cellphone gaming options. Some phones come with as many as ten popular and addictive games, including Tetris, blackjack, and various memory games. Nokia phones, popular with younger consumers, tend to have scores of games.

Chatboard

A few phones come with an Instant Message-like feature that allows users to send messages in near real time to other users who have the feature enabled at that moment. These phones often come with larger displays to accommodate increased message length.

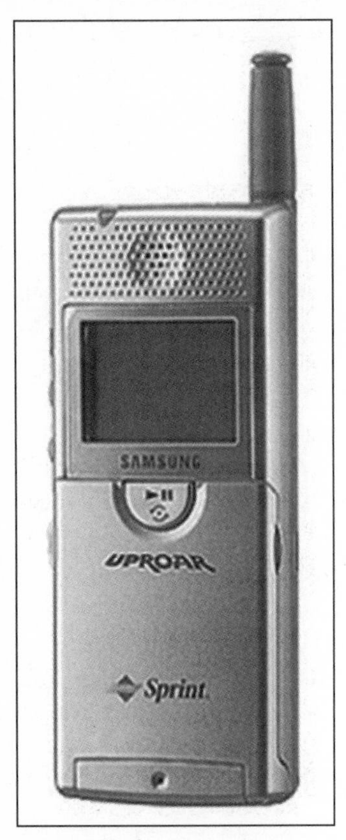

Figure 1-3: The Samsung Uproar M100 (photo courtesy of Samsung)

Java Applications

Some cellphone manufacturers are using Java to create custom user interfaces for enterprise and entertainment. Using JME2, a version of Java specifically for wireless developers, cellphone makers are creating Java-powered navigable application screens that can customize data and options available to users, as well as games with 3D interfaces. Some examples are the Motorola I85s and the Siemens SL42i (see Figure 1-4). Nokia has several JME2 phones in development.

Direct Connect

Direct Connect is a paired-phone option that enables you to establish a two-way radio connection with another paired-phone user. Direct Connect does not require users to be local. It uses a system called iDEN, a connection technology developed by Motorola to enable shared connections between workgroups. Direct Connect is now a feature offered most commonly on select Nextel phones.

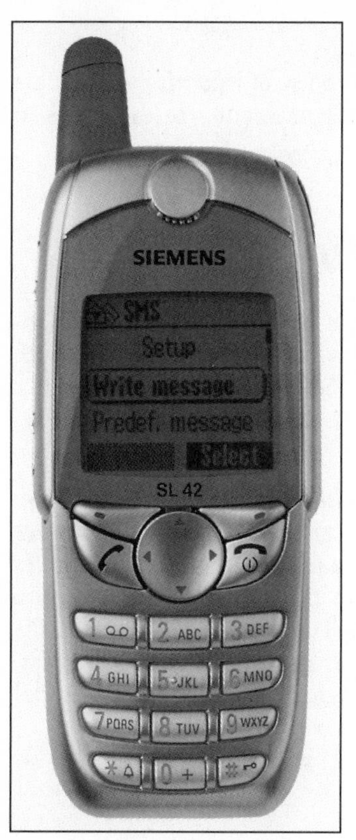

Figure 1-4: The Siemens SL42i Java-enabled phone (photo courtesy of Siemens AG)

Connectivity Options

Some phones provide connection ports to a PC, which allows users to access dial-up Internet service providers through their cellphone and thus use the mobile phone to provide Web access to a laptop or other device. (This topic is discussed in more detail later in this chapter.) Thus, the phone will have an RS-232 serial port connection or other PC-friendly interface. The cable port can also be used to back up your phone's contact information to a PC. Phones with communication ports include the Nextel i85s, Samsung 850, as well as the Nokia 3000 and 8000 series phones, and most newer Ericssons. Some phones — for example, the Siemens S40 — provide IRDA infrared connections, allowing data to be shared with PDAs. Of course, this feature would only be valuable on phones with PDA-like capabilities.

Other Sundry Features

Some phones include a password lock, which prevents unauthorized use of your phone, multiple language support on the LCD, and single-touch emergency dialing (which automatically dials an emergency number when a certain key is pressed).

Selecting the Right Phone Service

This section discusses factors regarding the selection of the right phone service options, as well as the right phone. Not all providers are created equal, and if the following features are important to you, you'll have to choose among available providers, selecting the best of the bunch.

Web and E-mail Access

Most cellphones and cellular services provide access to e-mail and Internet services of some type. Web-enabled cellphones aren't just for playing and shopping. They can give mobile employees live access to company data. For example, employees can interact with a database, updating customer information, entering orders, or changing a travel itinerary from the field. For these tasks, a device that can surf the Web like a PC is not a requirement. A small custom application that lets the mobile user remotely access necessary information fields does not require a large display. XML, Java, or other WAP scripting tools can easily be used to create small-screen applications that can pass data to and from an existing database.

The following sections talk about the different ways cellphones let you access the Web.

ACCESSING THE WIRELESS INTERNET

Some phones provide direct access to the wireless Internet, also known as WAP. You'll be browsing Web pages specifically designed for small-screened wireless devices (see Figure 1-5). These pages have been created using a derivative of HTML known as WML and WML script. Pages so designed will use mostly text and very few images.

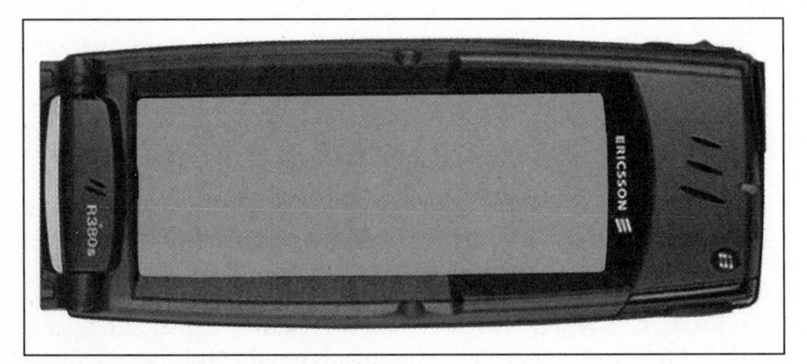

Figure 1-5: A WAP-enabled page view

Just because WAP-enabled sites are small doesn't mean that they aren't powerful. You can order online, check off preferences that will be remembered the next time you visit, and perform other types of e-commerce activities. Phones that can access these sites are called *WAP-enabled*. Since WML-designed sites are made for handheld wireless devices, these sites are usually pretty fast to load. Mindful that consumers are not apt to Web surf while the clock is ticking away precious cellphone minutes, some plans provide a special lower rate for Web usage.

SELECTIVE SITE ACCESS

Other phones provide access to a few select Web locations, such as Amazon, MapQuest, Yahoo, and other very proprietary sites. With such phones, you don't really "browse the Web" with your phone; rather, you choose from a list of preset online locations. Since these sites are especially designed to bring proprietary features to Web-enabled phones, they will usually load fairly fast.

WEB BROWSING WITH WEB CLIPPING

Other phones let you browse the Web fairly freely, employing a technology called Web clipping. This feature displays only a portion of the Web page — the portion deemed most relevant in terms of content. When viewed with Web clipping, pages with frames and lots of tables can look scrambled. Also, since viewing sites with Web clipping is more like random browsing than selecting options from a list, site loading can take a relatively long time.

CELLPHONE–TO–COMPUTER CABLING

Some cellphones provide a port connecting a cable to your laptop, PDA, or handheld computer. After installing the software and then establishing a dial-up connection on the cellphone, the computer can access the Internet as it would any other time. You won't be limited to the small display of your cellphone, and you can access any sites you like. Please note that the actual connection speed will be a fraction of what you are used to, even from a dial-up. Note also that you'll be using standard phone minutes from your cellphone's minute allotment, not reduced-cost Web-access minutes.

This section describes devices that simplify connectivity between cellphone and computer, thus making it easy to access the Internet wirelessly by combining both.

SOCKET'S DIGITAL PHONE CARDS

Digital Phone Cards (see Figure 1-6), available from Socket through Connect Globally (`www.connectglobally.com`), connect to your computer via a Compact Flash-to-PC adapter (many models are PCMCIA-compatible) and into your cellphone's data cable connection. Models are available for many phones and are also available for handheld computers as well as card-enabled laptops. The phones must have digital cellular service through GSM or CDMA.

OSITECH'S LAN OR WEB ACCESS CARDS

Ositech sells a series of cards that provide Internet or LAN access to laptop and handheld computers via a wireless cellphone connection. Many cards are available, providing digital and analog phones with a variety of network and Internet connections. These cards require that the computers are CardBus-compatible. Many laptops support CardBus technology, but it is not as universal as PCMCIA, so you should check with the laptop's manufacturer before purchasing one of these cards.

Figure 1-6: Socket's Digital Phone Card (photo courtesy of Socket Communications, Inc.)

E-Mail via Cellphone

Note that all the preceding wireless Internet access options give you access to e-mail. However, not all cellphone e-mail options are created equal. To begin with, the ability to access an e-mail account with your cellphone doesn't make it easier to type a response by using a cellphone keyboard. Secondly, in some cases, you'll still have to forward e-mail from your regular e-mail account to your cellphone account.

TYPING AN E-MAIL RESPONSE

If you think that you'll frequently use your cellphone to respond to e-mail, consider investing in a cellphone with a larger keyboard, for example, one of the cellphone/PDA hybrids such as the Handspring Treo or Sprint Kyocera. This option may especially be essential if your typed responses will be more than a sentence or two. Also, consider phones that allow you to create preset responses and send them with a couple of keystrokes.

FORWARDING AND SYNCHRONIZING E-MAIL

If your cellular service provides you with an e-mail account when you sign up for Web access, you can have e-mail forwarded from your regular account to this account using Microsoft Exchange or Outlook (depending on the type of account you have). Because of this forwarding, you'll notice a moderate lag in arrival when checking mail via your cellphone. When setting up via Exchange, if you use Inbox Assistant, you can discreetly forward the e-mail to your new, wireless account, which convinces the sender that you have responded from your normal e-mail account, which is a plus if you'd rather the whole world not know you were responding while on the road.

Other phone services let you set up at least one POP3 account, and most more than one. You can then forward as described earlier. Also, to provide complete synchronization between your mobile e-mail account and Outlook or Exchange on your desktop, create an IMAP e-mail account. IMAP is an e-mail standard developed as a subset of POP3 that enhances management of multiple e-mail addresses. For example, using IMAP, you can delete an e-mail from your mobile access point, and when prompted, affect the deletion on all other computers that access that account as well.

Event Notification via Cellphone

One Web feature that is growing more popular is update services, or specialized notification. When users sign up for Web access via cellphone, they can indicate a preference to receive news updates on topics they choose, notifications of product availability, and other special information. The key is that the user will not have to open a page or navigate a menu to obtain the information. A notification will appear on the cellphone display, accompanied by an audio alert. Thus, the information is said to be "pushed" — updates are automatically sent to the user as a result of stated preferences when the service was first initiated or preferences updated thereafter. Advertisers and news services of a proprietary nature greet this feature enthusiastically, because it guarantees that the users will see the notification simply by accessing their cellphones.

Before Signing Up

This section describes the various factors you should consider before purchasing a cellphone and signing up for a phone service.

DIGITAL, ANALOG, OR BOTH

In the rural United States, as well as on the open highway, digital coverage is still nonexistent or spotty. If you live in a small town or think that you'll be making calls from the middle of nowhere, you'll need an analog cellphone, or a phone that can switch between analog and digital (dual-mode). You'll also need a service that offers both modes of coverage without exorbitant roaming charges.

OVERSEAS USAGE

If you will be using your cellphone in Europe or Asia, you'll need a phone that can switch to the international GSM system. Although some areas of the U.S. are covered by GSM, the bandwidth is different. You'll need a phone with a physical switch that toggles to international GSM. Unless you purchase minutes from an overseas service provider once you arrive, expect to pay around $2 per minute for global roaming fees. Good world phone examples include the Ericsson T28, the Atcatel One-Touch, and the Siemens S40.

TALKTIME, WAP TIME, AND STANDBY TIME

Before buying a cellphone, find out how many hours you can use to talk or access the Internet before you must recharge. Also, find out how long you can leave the phone on standby without having to recharge.

CHARGE BY THE SECOND

Some cellular service providers have plans that charge by the second rather than rounding up to the nearest minute. If you make frequent, short phone calls from your cellphone, this plan will save you lots of money.

PREPAID VERSUS CONTRACT

In some situations, having a prepaid phone plan makes sense. If you are "credit-challenged," require an immediate account, or need a cellphone for a short time only (or have a short-term requirement for an additional account for an associate or family member), setting up a prepaid account can be the way to go. This option is especially helpful for a single instance of travel in which cellphone access is a requirement, either for one or several people.

Prepaid plans are great for situations that call for 24/7 accessibility, but not heavy cellphone usage.

Usage and Priorities

These days, cellphones can do more than simply make and receive calls. With so many phones available, it is easy to create a list of desirable features and find a phone that has most, if not all, of these features. The following categories may help you identify the features you find most appealing.

THE MINI-PHONE

As its name suggests, the mini-phone is very small. It fits nicely into a shirt pocket or small purse. To obtain ultimate compactness, you'll have to sacrifice keyboard and display size. This phone is not optimal for reading or writing lengthy e-mails.

THE MULTIMEDIA PHONE

The multimedia phone lets you play MP3s and tune in to local radio stations. You'll probably also be interested in its color display, the number of games included, and the Java and XML support for 3-D applications and advanced gaming.

THE CONNECTED PHONE

Connected phones provide many ways to connect with other devices, including serial ports for connecting your phone to a PC, laptop, or handheld computer, as well as an IRDA infrared port for sharing data with a PDA or other IRDA-enabled device.

THE WEB PHONE

A phone that emphasizes Internet connectivity will let you display WAP-enabled sites as well as Web clipping and will provide support for Java and XML applications that enhance the wireless Web experience. A generous screen should be provided for reading and responding to e-mail. A port for connecting the phone to a computer or PDA should also be provided.

THE TYPE-AND-TALK PHONE

The type-and-talk phone provides the ability to type and display many lines of text, utilize predictive text to minimize your keystrokes, include common preset messages, and let you fax text messages, depending on your available service. Some of these phones allow you to send and receive near-instant messages with others using a similarly enabled phone.

THE PDA PHONE

Phones that can fill in as personal digital assistants include features such as storing and sorting contact information, a scratchpad for jotting text notes, handwriting-to-text recognition, calendar and scheduling, and audio event notification.

THE INTERNATIONAL PHONE

The international phone allows you to make calls overseas without having to change phones. A switch lets you select GSM at 1800 MHz when traveling in Europe. Some phones also include a currency conversion feature for easy expense calculations when abroad.

Cellphone Deployment as an Enterprise Solution

In terms of business applications, other wireless devices offer more than cellphones in every case but one: voice-driven transactions and verbal-communication requirements. For maintaining and updating contact lists in the field, PDAs are superior. For portable business applications that require graphics or relatively large screens of information, handheld PCs are the natural solution. However, for any voice-driven application, or one that requires voice contact while simultaneously entering data, a Smartphone with a PDA OS such as Palm would work best. Rather than have field employees juggle both a cellphone and a PDA while talking and navigating data screens, deploy a Palm OS-based or similarly enabled Smartphone. The Ericsson R380 is a good example (see Figure 1-7).

Figure 1-7: The Ericsson R380 (photo courtesy of Telefonaktiebolaget L M Ericsson)

Choosing a Cellular Service for Business

Selecting a cellphone provider for business usage requires you to evaluate the geographical area that needs to be covered, the types of services offered, and how those services will be improved in the near future. If cellphone deployment to empower mobile employees is being planned, keep the following in mind:

- ✓ Compare the area being covered with the area where the mobile employees will spend the bulk of their time. Do they overlap for the most part?
- ✓ If Web access from a mobile phone needs to occur frequently in your business, make sure that your service offers high enough bandwidth to make it worth everyone's while. Some services offer only a 9.6 Kbps connection. If mobile employees will have to upload or download lots of information from the field, a slow connection may not be adequate.
- ✓ Beyond bandwidth and geography, look at the types of wireless services being offered. Sending voice and data simultaneously requires specific technologies, as does WAP access or fax transmission.
- ✓ All wireless service providers are gearing up to provide more, as various new technologies become available. Higher bandwidth, better geographical coverage, and more service types will be offered. Find out what exactly will be implemented as well as estimated rollout dates.

Pagers and Wireless Messaging Devices

Walkie-talkies, CB radios, and pagers were the first examples of popular wireless communication technology. In fact, they all share the same inventor, Al Gross, who patented the pager in 1949. The first pagers that became ubiquitous on every teenage boy's belt buckle in the 70s and 80s were fairly simple devices. They allowed a telephonic device to send enough alphanumeric characters to comprise a phone number. The signal, sent over the analog public telephone system, would be received by the wireless receiver — the pager. The signal would be demodulated and then the alphanumeric digits displayed on a small LCD. The recipient would then find a phone and return the page. Clever users of these early one-way pagers would devise codes that could communicate important messages at a glance. For example, the message "777" could inform a physician that he needed to "get to the hospital fast — you have to operate."

Two-Way Paging

Looking at today's two-way digital pagers, you scarcely recognize their humble origins. Although you can still purchase simple pagers, activate them, and receive beeps, today's modern pagers are quite different. They are actually two-way e-mail terminals, complete with keyboards and enough memory to maintain huge contact lists and run a basic e-mail synchronization program remotely with your PC.

Pagers are great for those who cannot afford or trust themselves with cellphones (a good pager service runs around $12 per month) but still need to be accessible. However, pagers are advantageous in other ways. For example, they can be used in conjunction with Global Positioning System devices to remotely track vehicles from a dispatch center.

In between one-way and two-way pager technologies are pagers that notify the sender that the page has been received. When the recipient presses the Receive alert, a signal is routed to the sender, indicating reception.

Messaging Devices

As of this writing, the most popular pagers are made by Motorola and Research in Motion, although they are more accurately referred to as messaging devices. For example, Motorola's popular Advisor Elite (see Figure 1-8) is a one-way alphanumeric pager that also lets you receive e-mail. If your pager service is so enabled, people can e-mail you from any e-mail-enabled device just by indicating `@airtouch.net` (or some other service) at the end of the phone number. For example, the sender would type an e-mail and then send it to `832-345-8321@verizon.net`. The pager would display the e-mail text. The Advisor Elite displays four lines of text and will zoom to two lines of text for easier reading. You can scroll to read e-mail messages longer than four lines. Also, the Advisor Elite has 34 message slots and a 30,000-character memory. Since you cannot send messages from a one-way pager, keyboards are not provided.

Depending on the provider you choose, you can receive news notifications from various media services as well, including sports and entertainment announcements.

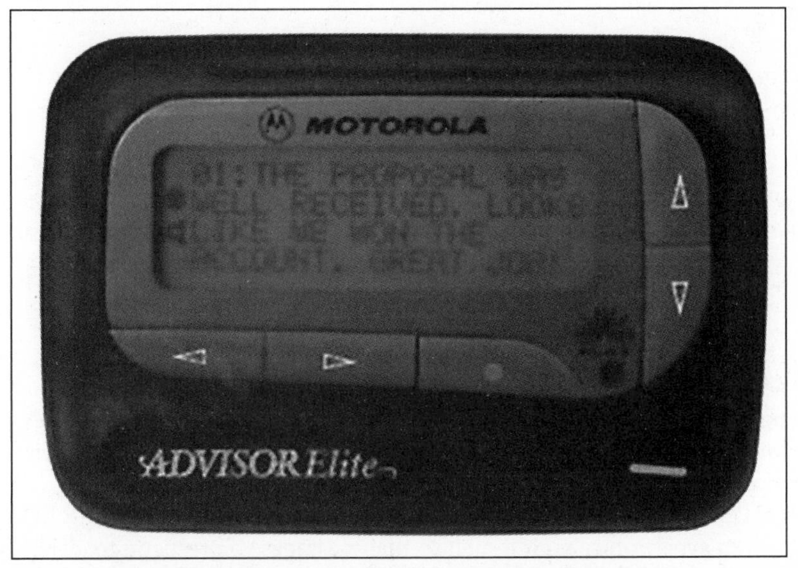

Figure 1-8: The Motorola Advisor Elite (photo courtesy of Motorola, Inc.)

Two-way messaging devices such as the Motorola Talkabout (see Figure 1-9) do come with keyboards. The units are larger than one-way pagers but still very portable, measuring about 3.1 × 2.1 × .9 inches and weighing 4.2 ounces. The T900 Talkabout keyboard, just like RIM's Interactive 950 pager, is easy to type with, considering its small size. Like most modern pagers, you can be notified of a message by beep, vibration, or both. You can also receive pages without any notification. They'll still "go through," but in complete silence.

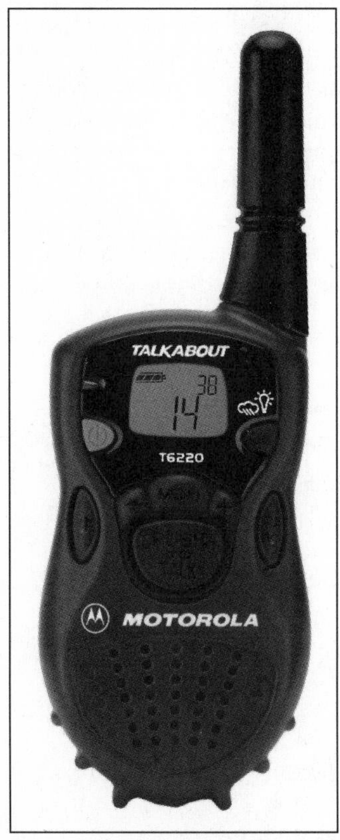

Figure 1-9: The Motorola Talkabout (photo courtesy of Motorola, Inc.)

Wireless E-Mail Management

In essence, carrying around a device such as the Motorola Talkabout or RIM Interactive 950 pager (see Figure 1-10) is like having instant e-mail access everywhere you go. Depending on the service provider you choose, you can either send and receive a limited number of messages per month or forward e-mail from your Outlook or Exchange inbox to your portable unit. Using the RIM Interactive pager, you can even synchronize your e-mail so that messages you respond to, delete, or keep as new will appear as such from your other e-mail access points.

Figure 1-10: The RIM Wireless Interactive Pagers (photo courtesy of Research In Motion Limited)

Although the Motorola T900 Talkabout and RIM Interactive 950 pager let you maintain and sort very large contact lists and provide timed notification for alarms you set, they do not perform scheduling or other PDA-like tasks. They are sophisticated messaging devices, complete with predictive text features and various auto-correct functions.

The RIM BlackBerry E-Mail Solution

The RIM BlackBerry is presented not so much as a personal messaging device, but as an always-on e-mail and notification solution for business professionals. Some of the nation's largest telecommunications service providers offer exclusive RIM BlackBerry services to businesses that deploy BlackBerry units to their on-the-go employees. Currently, BlackBerry comes in two editions: BlackBerry Enterprise Edition, for corporate environments, and the Internet Edition, for wireless access to e-mail.

Three main features make the BlackBerry attractive as a business solution:

- ✓ The BlackBerry is designed to be always on. You never need to press a button to know you've got e-mail. You'll be notified instantly using the notification method of your choice.
- ✓ BlackBerry provides seamless service for employees in the office, on the factory floor, anywhere else "in the building," or hundreds of miles away. E-mail service runs uninterrupted between LAN and WAN. Plus, many providers offer customizable solutions to companies that want to use BlackBerry devices to integrate e-mail, voicemail, interoffice notification, and faxes into one portable notification device. Setting up BlackBerry software on each BlackBerry user's office PC does this on a corporate-wide level. This software integrates each BlackBerry unit with the corporate Microsoft Exchange or Lotus Domino server configuration.

- ✓ BlackBerry is very secure, providing triple DES encryption. (Triple data encryption standard is an encryption method that encrypts a message three times before sending it.) E-mail messages remain encrypted at all points during the transmission process. Additionally, BlackBerry provides scheduling and calendar support that can be integrated into Outlook for appointment notification.

Summary

You now know enough about cellphones and pagers to make an informed purchase, as well as some options for equipping your phone with data and voice transmission technology. In the next chapter, we discuss personal digital assistants and the Palm OS.

Chapter 2

The Wireless PDA

In This Chapter

The following topics are covered in this chapter:

- ✓ Defining the personal digital assistant
- ✓ The PDA as a unique wireless device
- ✓ PDA product lines
- ✓ The PDA in daily life
- ✓ Going wireless with a PDA
- ✓ PDA solutions

What Is a Personal Digital Assistant?

Broadly speaking, a PDA, or personal digital assistant, is a handheld digital organizer that stores contact information and allows notetaking of some sort. For notetaking, many PDAs use Graffiti, a shorthand alphabet developed by Palm for generating text on a Palm screen with a stylus. PDAs are also called by other names, for example, Palm Pilot or Organizer. Pocket PCs will be dealt with in Chapter 3.

The three most significant PDA manufacturers at the moment are Palm, Handspring, and Handera. All PDAs made by these manufacturers run on the Palm OS, an operating system developed specifically for PDAs. The Palm OS makes it possible to create software and develop features for all PDAs, rather than one or two models. Palm OS–based PDAs are the focus of this chapter.

We'll also discuss hybrid PDA/phones, devices that combine a PDA and cellphone in one clever unit. Not all PDA/phones use the Palm OS. The Ericsson series and Psion, both manufactured in Europe, use the EPOC OS, developed by Symbian and discussed in detail in Chapter 4.

This chapter describes the technology and use of the PDA as a wireless communication device. You'll learn about the hardware and software that is required to go wireless with your PDA and also how to deploy wireless PDAs as viable business solutions. This chapter also describes software programs that enhance the sharing of wireless information between PDAs and other devices.

The PDA's Unique Niche

The PDA occupies a unique place in the digital marketplace. It seeks a perfect marriage between ergonomics and capacity. "Small" is convenient, but "too small" becomes impractical, and essential features get left out in the name of portability. For example, how tiny can you make the buttons before users have trouble pushing them? How small can you make the fonts before users can't read them? How complex of an application can you actually run on a 2-inch screen?

PDAs are facing competition from the Pocket PC. However, the Pocket PC's mandate is different from that of the PDA. It exists to miniaturize the features of a laptop, provide multimedia capability while on the go, and carry a decent set of desktop applications for work and play. PDAs are essentially information managers, however advanced they become. Still, in the face of competition from more powerful Pocket PCs, PDA manufacturers emphasize miniaturization and extreme portability. To answer the challenge of Pocket PCs, PDAs get larger and lose a little ergonomics as a result.

Users want wireless connectivity, MP3 playback, more storage space, and a bigger, full-color screen. However, a bulky PDA would defeat its own purpose: It would get left at home. If PDAs become too large, they will no longer be the incredibly lightweight and essential contact-management devices that we all love. They'll be thought of as underpowered laptops and get left in a drawer to rot.

Thus, the PDA is being pushed in various directions, trying to keep the microdevice enthusiasts happy on the one hand while competing with handheld PCs on the other. Finding an essential feature set that can still fit into the palm of one's hand is the name of the game. With wireless capabilities, the PDA offers a tiny but substantial window on the world.

The PDA's Wireless Future

Wireless communication technology is an integral part to a PDA's capabilities. A reliable and fast wireless connection could ensure the PDA's future. By marrying the small size of PDAs with the potential of unlimited wireless communication, PDAs can become the new "PC," the device that revolutionizes the world.

Sticking to Practicality

As technology advances, you will consistently see PDAs with added features. However, your decision on what handheld to buy should be based on the types of tasks for which you will use your PDA rather than on a desire for the newest add-on or faddish attachment. In general, PDAs perform the following functions:

- ✓ **Contact management:** Newer PDAs let you e-mail, or even telephone, contacts without reaching for a different device. The PDA's primary "reason for being" is to manage information about — and share information with — people you know.
- ✓ **Information retrieval:** Professionals of all types turn to PDAs to store and retrieve information. For example, doctors use PDAs to look up drug dosages and contraindications, locate laboratory data on a patient, and search medical journal archives. Today's PDAs can hold more than 16MB of data, which means you could easily carry an entire medical library in your pocket.
- ✓ **Daily finance management:** Most people use PDAs to keep track of daily expenses. With a PDA, you can quickly record a spontaneous purchase or incurred expense.

- **Mini reader:** Thousands of books are available for PDAs, and more are being added all the time. Reading a book on a PDA is surprisingly easy, thanks to the sharp, crisp lettering of most of today's units. In addition, many PDA readers have auto-scrolling, allowing you to read your mini-book "hands free."
- **Web portal:** Few relish the experience of "surfing the Web" on a PDA. The small screen size and slow Internet connection would try anyone's patience. But searching out specific information online — especially if you have a good idea where to find it — is quite doable. The Palm VIIx even knows where you are when you log on. Use the Palm VIIx to look for the nearest movie theaters, and it will know where to look.
- **Network data access:** Mobile employees of large or small companies can share information via PDA over secure network connections. PDAs can be synchronized centrally so that information updated on a central server won't need to be deployed to each employee's desktop computer.
- **Vehicle dispatch and locator:** Companies with mobile fleets can use PDAs to monitor vehicle locations, dispatch, and job reporting. Products can be tracked throughout the delivery cycle, improving customer service with timely information.
- **Mobile conferencing:** A PDA user can participate in or organize a conference with multiple participants via the Web. The mobile PDA user can display a PowerPoint presentation, draw diagrams, and distribute images viewable by all Web conference participants.

Current PDA Product Lines

The following section discusses some of the current PDA product lines and major classifications, as well as hybrid PDA/phone devices.

The Palm

The Palm PDA is the most popular type of PDA. In fact, the tremendous popularity of Palm has spawned the development of literally thousands of applications, ranging from document converters to e-mail enhancement to games.

All Palm handhelds include basic features such as Address Book, Date Book, To Do List, Memo Pad, Expense 1, and so on. These features are conveniently laid out, and even with the addition of new features, the units continue to be sleek and small. The battery life has remained high, and rather than load the devices with bulky options, Palm prefers to keep them portable. Here is a quick look at the current Palm model lineup.

- Palm's legacy handheld, the III series (see Figure 2-1), has huge numbers of devotees. Palm III users load up their handhelds with e-books, software of all types, modems, and portable keyboards. Popular right now is the Palm IIIc — Palm's first color unit, with 256 colors. The sturdy and lightweight Palm Vx seeks to retain Palm's traditional strengths of portability and comfortable use. However, like other newer Palms, information can be "beamed" via Infrared port to other units, and software upgrades are quickly executed with flash memory.

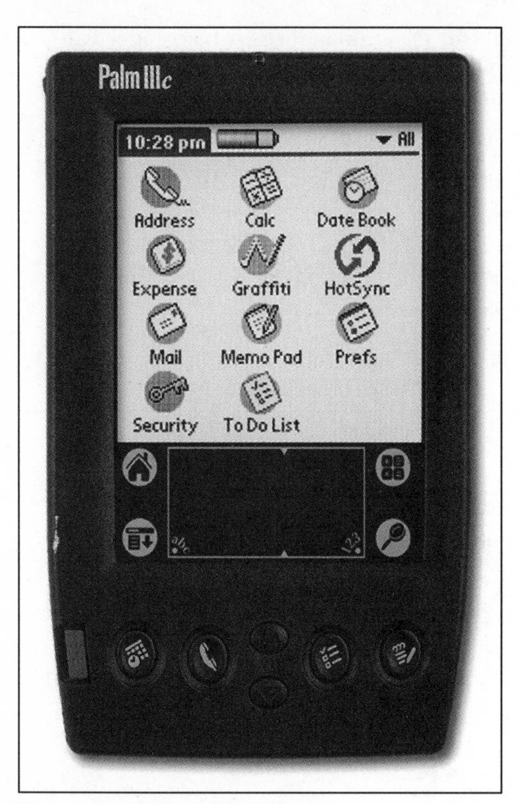

Figure 2-1: The Venerable Palm IIIc
(photo courtesy of Palm, Inc.)

- ✓ Devices in the Palm m100 series are the "hip" models, with interchangeable color faceplates and an optional MP3 player. Encouraging self-expression, the Palm m100 series lets you write "sticky notes" directly on the screen, using your own handwriting. At 2MB, the Palm m100 provides less RAM than any current Palm model. All other models provide at least 8MB. Devices in the Palm m500 series are the first to provide expansion slots (Secure Digital/MultiMediaCard slots) for adding peripheral devices such as digital cameras and modems.
- ✓ The Palm m500 series is deployed on the Palm 4.0 operating system, with increased application speed and increased battery life. In addition, the m500 series provides a USB connection for synchronization. The Palm m505 was Palm's first "really good" color unit, with a high-resolution 65,000-color display. The m500 and m505 both come with lots of application software, such as DataViz Documents to Go and the Palm Reader for e-books. With a standard 8MBRAM for applications, you can upgrade an m500 or m505 device to 64MB by using a Compact Flash memory card.

✓ The Palm VIIx (see Figure 2-2) is the first PDA to integrate wireless Internet access, using a wireless transceiver built specifically to work with Palm's Web portal, Palm.net, and Web clipping application files. With the Palm VIIx, you can send and receive e-mail from most e-mail account types. Even though you are limited to access via Palm.net, the Palm VIIx provides a remarkably fast Internet connection, in contrast to several other Internet connection methods used by PDAs.

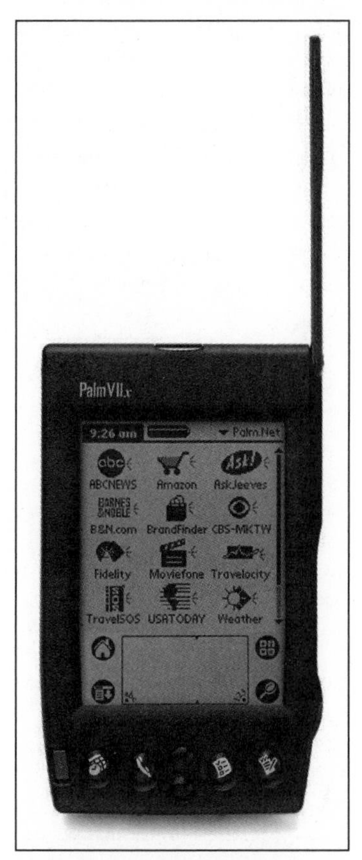

Figure 2-2: The Palm VII, the first PDA to offer integrated wireless Internet access (photo courtesy of Palm, Inc.)

All newer Palm models include basic software, such as Chapura's PocketMirror, for synchronizing Microsoft Outlook with your Palm contacts and appointments. Although very few Palm models come with integrated Web access, all include software that help bring the Web to your Palm. AvantGo, for example, lets you specify sites that you want available on your Palm. When you synchronize your Palm with your PC, the AvantGo program will download content from the requested Web sites to your Palm, enabling you to access your information from those Web sites any time you like. Your Palm may also include the Mobile Internet Kit, which lets you access e-mail and the Internet by using a data-enabled mobile phone.

Handspring

Another company with a product line large enough to warrant a comprehensive discussion is Handspring. Handspring uses the Palm OS, so all software that works on Palm handhelds will also work with Handspring devices.

Handspring handhelds all come with expansion slots. Equally important, you can find plenty of optional modules to fill the slots, including wireless communication modules, MP3 players, e-books, memory and storage, productivity enhancement modules, and even enhanced GPS modules.

All Handspring handhelds connect to your computer using USB, for fast synchronization, and all have an Infrared port for "beaming" data and application transfers to other IR-enabled devices. All Handspring handhelds come with at least 8MB of RAM, and expansion slots can be used to increase memory capacity. Currently, the Handspring Visor Prism (see Figure 2-3) is Handspring's only color unit (16-bit, 65,000 colors), while the Visor Pro offers the most memory off the shelf (16MB). The lightest Handspring handheld, the Visor Edge, weighs in at 4.8 ounces. Palm enthusiasts point out that Handspring units are a bit bulkier, and not as ergonomically streamlined, as the Palm. However, Handspring points out that, dollar for dollar, their devices provide more memory and features, as well as expansion slots with dozens of modules that are ready to go.

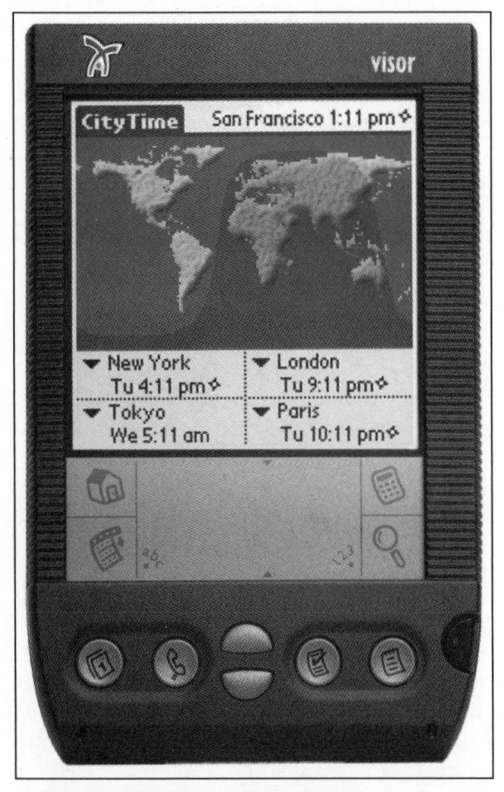

Figure 2-3: Handspring's first color PDA, the Visor Prism (photo courtesy of Handspring, Inc.)

Sony CLIE

The Sony CLIE PEG N710 PDA is remarkable due to its color display (see Figure 2-4). The CLIE PEG N710 has a reflective TFT screen with a resolution of 320 × 320, currently twice that of other color models. It has 8MB of memory, upgradeable to 64MB, and an MP3 player. The device also has 4MB of flash memory for easy upgrades and the power to run large applications. With its superior color, roomy memory and expandability, Sony hopes to position the CLIE models in a unique niche between other PDAs and handheld PCs.

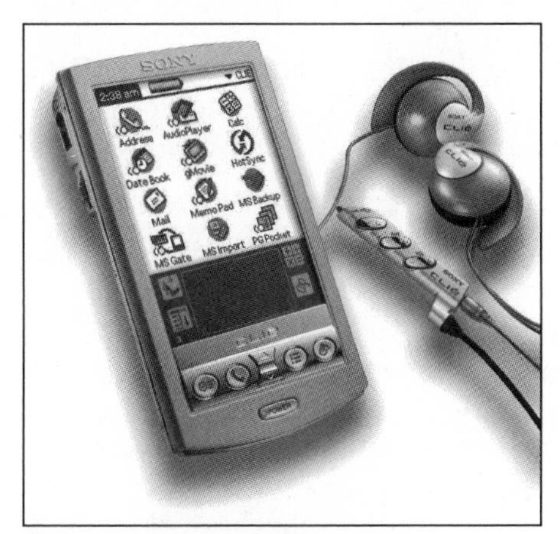

Figure 2-4: The Sony CLIE PEG N710 has a 65,000-color display at double the resolution of most other PDAs (photo courtesy of Sony Electronics).

PDA/Phones

One contender for the future of PDAs is the hybrid PDA/phone. The marriage of PDA to cellphone is an obvious one. The phone doubles as a Web-access device. You can look up any contact in your PDA and ring them with a single push of a button. The PDA's larger screen is much better for using e-mail and viewing Web pages. Provide a larger alphanumeric pad for text entry, and you've pretty much got yourself the long-sought Swiss Army Knife of mobile wireless communication. Some PDA/phones provide a rotating PDA screen, so pages can be viewed in landscape mode — useful for viewing spreadsheet columns as well as Web pages.

Indeed, PDA/phone hybrids tend to be both very good PDAs and very good phones. However, in the past, these devices have stretched the boundaries of portability. The early PDA/phones were large and bulky, and battery life suffered greatly. Also, as a general rule, PDA/phone hybrids are not upgrade-friendly. They come packed with features, but you cannot easily upgrade the operating system or add memory. Some limit the type of Web access you have. For example, the Kyocera QCP-6035 does not have a built-in WAP browser.

The following sections offer a look at a few of the PDA/phone hybrids available.

HANDSPRING VISORPHONE

One PDA/phone hybrid that overcomes some of these limitations is the VisorPhone/Visor Prism combo. You can use expander modules with the Visor Prism, and the VisorPhone supports WAP and the various microbrowsers such as Blazer.

KYOCERA SMARTPHONE

The Kyocera series provides standard PDA features such as Address and Date Book, Memo Pad, and To Do List. Since it runs on the Palm OS, the Kyocera can use any of the Palm applications that work on other Palm OS devices. In addition, it supports Infrared data transfer, provides voice dialing and voice memo features, and comes with a speakerphone, which enables you to talk on the phone while using the PDA functions. An input device on the left side of the phone, called the Shuttle, is a four-way "rocker" button for single-handed screen navigation. The navigation button, sometimes called a Jog button, functions a bit like a miniature version of a laptop mouse button.

HANDSPRING TREO

The Handspring's Visor Treo PDA/phone has a generous number of features, and is very compact (see Figure 2-5). The phone has a large 160 × 160 pixel screen, a USB connection port for synchronization, and a Jog button for convenient one-handed navigation. The Visor Treo comes in two models: one with a Graffiti writing area for text entry with a stylus, and the other with a built-in keyboard. The Treo has 16MB of memory and uses the Palm OS 3.52H4. It does not have a Springboard Expansion module or flash memory, so users will not be able to upgrade the operating system or add new hardware to the phone. The lack of expandability, though, helps keep the size down. The Treo is smaller than most wallets, which is quite an accomplishment for a hybrid PDA/phone. Handspring is betting that users looking for expandable devices will consider other Visor handhelds.

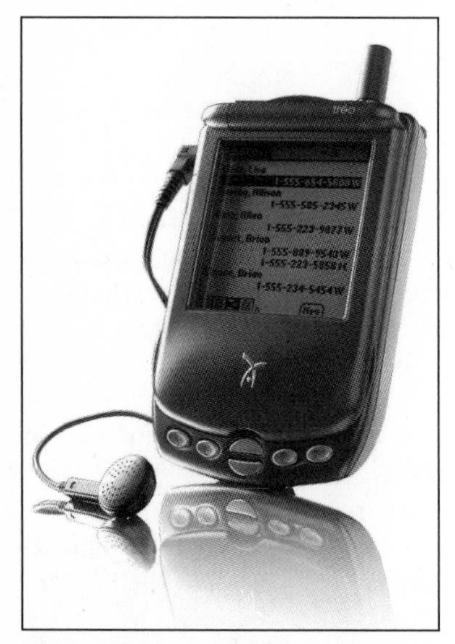

Figure 2-5: The Handspring Treo is full-featured and compact but offers few upgrade options (photo courtesy of Handspring, Inc.).

The Modern PDA at Work

This book is not interested so much in a PDA, but rather in a PDA all dressed up and ready to go. What can you do with such a device? This section describes a typical day in the life of someone who owns a PDA that is enabled with advanced Web and network access and synchronization:

- ✓ At 10:00 a.m., away from the office, you round up a list of contacts on your PDA and set up a conference call. You use your PDA to draw a quick diagram pertinent to the call and upload it to a Web site viewable by all conference call participants.
- ✓ At 11:00 a.m., you realize that you have an important contact on your desktop computer. You access your network and download the contact to your PDA.
- ✓ At 1:00 p.m., you head towards an appointment at the Starbucks "nearest downtown." You use your PDA to confirm the address of that particular Starbucks.
- ✓ At 3:00 p.m., you upload your meeting notes from your PDA to your secretary's Outlook inbox so that he can type them up and present them to the boss later that day.
- ✓ At 4:00 p.m., you stop by the office and bump into the IT director, who says it's time to synchronize your PDA with the new software upgrades. You do so by leaving your PDA at the edge of your desk for a few minutes, within range of a Bluetooth wireless connection.
- ✓ At 6:00 p.m., you stop by the gym and work out on the treadmill for a while. You open your PDA reader to your last bookmark of a new novel, set the PDA on auto-scroll, and read while you work out.
- ✓ At 10:00 p.m., you HotSync your PDA to your PC and simultaneously recharge the PDA's battery. You are also synchronizing with AvantGo so that information from your personal favorite Web sites will be saved on your PDA and browsed at your convenience.

Getting Ready for Wireless

A wide variety of PDA brands and models is available in today's market. Besides a wireless connection, what else does a PDA need before it can go wireless and be unleashed on the world?

- ✓ An expansion slot of some kind.
- ✓ A modem. Note that modems for PDAs are model-specific. When researching your modem, make sure you buy one that will be compatible with your specific model.
- ✓ Adequate storage space.
- ✓ A modem that you are happy with because the cost of the modem could indeed exceed the cost of the PDA.

Finally, make sure that you get a PDA you'll want to keep for a while. Since modems are model-specific, you'll be stuck with both.

Going Wireless with Your PDA

The following section describes the different technologies available for accessing the Web and using e-mail with a PDA.

The Palm VIIx/Palm.net Option

One option is to purchase a Palm VIIx or similar device that has out-of-box integrated Internet access. The Palm VIIx does not have a modem; rather, it has a wireless transceiver that's specifically designed to work with the Palm.net service and to view Web clipping application files. Palm developed Web clippings (WCAs) to provide users with quick snippets of Web pages with essential information. When a page is accessed, narrow and specific information options are displayed. You'll see, for example, a small form for making a purchase, or a list of nearby movie theaters and their schedules. The Web clippings load fairly quickly on the Palm VIIx. Currently, hundreds of Web clipping sites exist, and the list is always growing. The Palm VIIx also provides localized information as well as e-mail access.

With the Palm VIIx, you won't be able to browse sites via microbrowsers, such as Blazer, or use AvantGo to synchronize Web information from your desktop. Also, since the Palm VIIx's transceiver requires the BellSouth Mobitex network, it currently only provides access in major metropolitan areas in the United States. However, devices similar to the Palm VIIx may expand on these options. Also, software solutions that expand Internet access on the Palm VIIx exist, such as Digital Path's DPWeb (discussed later in the chapter).

The PDA/Phone Option

You can obtain Internet access by purchasing a hybrid PDA/phone. This choice offers immediate Internet connectivity and completely integrates PDA applications with your cellphone. For example, hybrid PDA/phones let you locate contacts on your PDA and call them in a single-handed operation. PDA/phones are usually sold with complete access plans, ensuring decent phone service, e-mail access, and a Web connection.

The Modem/Service Plan Bundle Option

Another option is to purchase a wireless modem for your PDA — for example, the Minstrel modem (see Figure 2-6), Sierra Wireless AirCard, or the Novatel Sage Wireless series. Handspring also offers Springboard modem modules for Visor expansion slots. These modems provide complete Internet access, allowing you to view WAP pages and Web clippings, and to browse with Blazer and other microbrowsers. Most add-on modems for PDAs use Cellular Digital Packet Data (CDPD) technology.

CDPD is IP-capable, and taps into the unused capacity of the AMPS system in the United States. CDPD uses packet switching, and thus, the user is theoretically charged by the amount of data downloaded rather than connection time. About 80 percent of the U.S. is covered under CDPD, and probably the best way to purchase a CDPD modem for your PDA is to select a provider. Most providers offer package deals that include the modem and a monthly access plan. Providers include OmniSky/Earthlink, GoAmerica, AT&T Wireless, and Verizon. CDPD service provides theoretical data transfer rates of 19.2 Kbps, the actual results depending on network conditions.

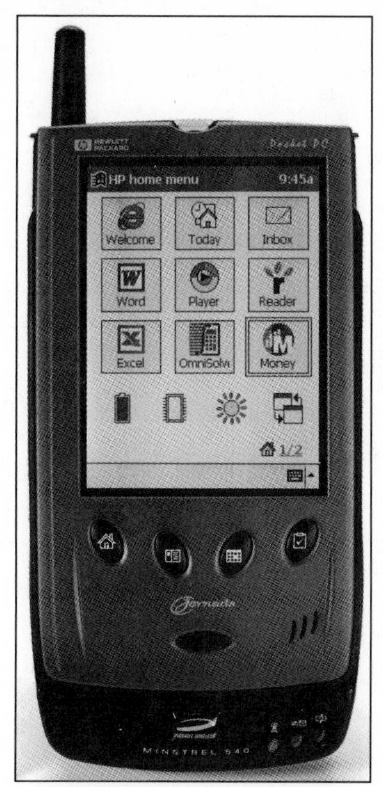

Figure 2-6: The Minstrel Wireless Modem (photo courtesy of Novatel Wireless, Inc.)

The Wireless Network Option

If you have access to a corporate network, you can wirelessly connect to it with your PDA by using a wireless network card. Currently, the most popular cards are part of the Sierra AirCard series (see Figure 2-7). Depending on how your network is configured would presumably give you Internet and e-mail access as well. This is a great wireless solution if you are on the road and require access to your network and e-mail services back at the office.

You can also wirelessly access the Internet with your PDA by using a data-enabled cellphone and special connector cable. Products that offer this option include Teladapt's Palm DirectConnect Null Modem Adapters and Ositech's wireless cards and cables. You simply connect and dial up your ISP on your cellphone. However, your cellular service provider may not offer data connections on your current plan, so you'll have to check first. Note that the connection speed will be very slow, around 8 Kbps or less. Also, with this connection method, you'll be charged for the minute usage on your cellphone. Connector card and cables for cellphone-to-PDA connections are very phone-specific. Before purchasing, make sure that your cellphone model is data-enabled.

Figure 2-7: The Sierra AirCard (photo courtesy of Sierra Wireless, Inc.)

PDA Power

This section discusses the various ways that PDA power is expanding. The number of applications for the PDA grows exponentially every year. PDA application development tools are abundant. The PDA's platform is fairly simple compared to that of the PC or Mac OS. The footprint for a PDA application is small, and developing a successful application that runs on the Palm OS is much less daunting than developing one for desktop computers. Also, the PDA's capacity for wireless communication, as well as expansion slots for an ever-growing list of devices, has broadened the PDA's potential.

- ✓ For individual PDA users, new software tools have been developed that can change the way you use your PDA to view the Web, work with e-mail and messaging, edit and view documents of all types, set alarms and alerts, and integrate your PDA data with your desktop applications.
- ✓ As an IT head or communications officer for your company, you may be responsible for deploying PDAs as a tool to keep mobile employees integrated with the rest of your company's workforce. You'll want to know whether employees' desktop applications can also run on PDAs, and whether field employees can securely update company data by using wireless PDA connections. You'll probably want to know if you can get a snapshot of all existing PDAs, their locations, and the data they currently contain, as well as network-wide synchronization and updates. We'll look at software and other solutions that would be helpful to both individual and corporate users.

How Do I Find PDA Applications?

This section is somewhat software-intensive. You may want to know where to find PDA applications and learn more information about them.

The following Web sites are excellent resources for locating Palm OS software and for obtaining related information:

- ✓ www.palmblvd.com/index.html
- ✓ www.palmgear.com
- ✓ www.nettechinc.com/palm.htm (Palm resources for lawyers)
- ✓ http://freewarepalm.net
- ✓ www.zdnet.com/downloads/pilotsoftware
- ✓ www.palmpilotware.com/
- ✓ www.palm.com
- ✓ www.purepalm.com/site/
- ✓ www.pdamd.com/vertical/home.xml (Palm resources for doctors)
- ✓ www.palmpower.com/ (*PalmPower* magazine)
- ✓ www.edteck.com/palm/ (Palm resources for professionals, especially educators)

PDA Software Tips

The installation of Palm applications is very easy. However, before downloading and installing a PDA application, keep the following in mind:

- ✓ The online applications you'll find at popular handheld sites will generally not be written for a specific device. Rather, each application indicates the minimum Palm OS version required to run that application. Whatever device you are using, your Palm OS must meet that minimum. For example, an application written for Palm OS 3.5 cannot be run on a handheld using OS 3.0.
- ✓ PDA software will not only have minimum requirements for your handheld but also for the host desktop computer, if applicable. For example, a program that organizes file transfer to your desktop will specify which desktop OSs are supported.
- ✓ Note the application's port requirements, if any. Some applications are designed specifically for IR ports, others for USB, and others for serial-enabled PDAs.
- ✓ Proportional to desktop computers, handhelds do not have much storage space for applications and files. Palm application developers know this, and they design their programs accordingly. Nonetheless, keep track of your remaining amount of storage and note the size of the applications you are downloading. Note both the size of the downloaded file — the archived application — as well as the application size once it is installed.

Choosing Web Viewing and Access Options

This section discusses how to determine your PDA's Web viewing options. Some PDA wireless configurations limit the type of sites available to you. Workarounds will be discussed, as well as the various options for taking more control of your PDA wireless experience.

Web Browsing

The OmniSky Internet service (recently purchased by Earthlink) lets you access any Web site you like, although you are primarily directed to sites optimized for handhelds. For those who really want to surf the Web from their PDA — not simply view Web clippings or be limited to WAP-enabled sites — the Blazer microbrowser provides access to a wide range of Web sites. You can bookmark your sites and set your own default home page as well. The Blazer microbrowser ships with Handspring PDAs and PDA/phones and can be easily installed. However, that does not mean that your wireless access provider supports full HTML Web browsing. If that issue is important to you, ask specifically before signing up. For more information about the OmniSky/Earthlink service, visit their Web site at `www.omnisky.com`.

Plenty of microbrowsers are in the works from Nokia and other companies. Most of these apply special tags or subsets of HTML, and thus, only pages optimized for that microbrowser's environment will be viewed correctly. However, those who have tried to view full HTML pages on a PDA can see why special configuration for PDA viewing is in order.

Getting Around the Web Clipping Limitation

For PDA access services that provide Web clipping only, one option is the Digital Path's DPWeb software application. DPWeb is an enhanced browsing environment that provides access to a limited number of sites for the Palm VIIx/Palm.net configuration and other wireless PDA devices. Rather than switch from a Palm.net Web clipping environment to a full CDPD-modem OmniSky environment, you can use DPWeb. The browser opens up "the whole Web" to your PDA, which was previously limited to Web clipping sites. You can bookmark sites and save Favorites as well. Many users enthusiastically endorse DPWeb since it readily expands the types of sites that are viewable in limited microbrowser environments.

Another program that allows you to build more of a "personal" portal on the Web is an Internet access kit, known as Mylo, which provides a portal for personalized Web content. This configuration, currently available only for the Sony CLIE series, lets you set up a home page with just the content you want delivered on a daily basis. This page could include top news stories, weather, sports, stocks, and so on.

WAP-Enabled Sites

Viewing WAP-enabled sites is also an issue for some PDAs. PDAs configured with the right software and service provider can view sites developed for the Wireless Application Protocol, or WAP. WAP-enabled sites are complete Web sites developed exclusively for microbrowser viewing (see Figure 2-8). They will vary greatly in appearance, depending on the microbrowser and service provider you are using. Sites optimized for PDA viewing with WAP are best viewed with browsers, and service providers recognize and can work with the Wireless Application Protocol stack. The

WAP stack is comprised of WML (the Wireless Markup Language), a bitmap image format (WBMP), WMLScript, and the Wireless Telephony Application. WAP sites are accessed through the WAP gateway and, thus, load fast and look great on a handheld. You actually get the impression you are viewing a site developed specifically for your device, rather than making do with scraps. Read up on how your browser and service provider display WAP sites and find out what your options are before making a selection, if you do have a choice.

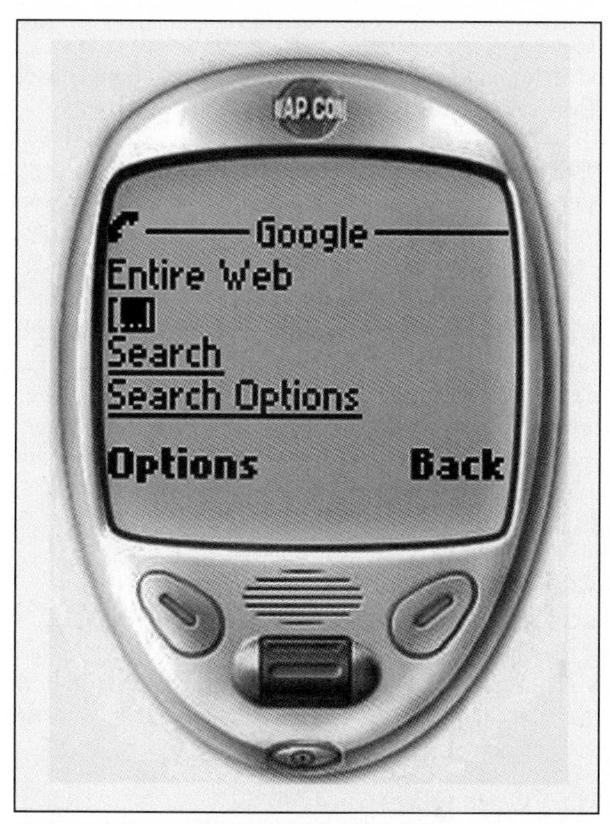

Figure 2-8: WAP-enabled browsing of the Google search engine

Integrating Web Content

If you want to easily organize and integrate Web content into your PDA applications, look into iKnapsack. iKnapsack is an Internet content management program. Items you view on the Web can be quickly integrated into a relevant application on your PDA. Addresses and dates can be added directly to your organizer. Online purchase amounts are saved into your account application. Web content you view can be stored as a memo or as a To Do item. Multimedia data can be integrated as well. Items you view on the Web can instantly become useful information, integrated into your PDA applications. Also see Eudora's Internet Suite, which synchronizes saved "Favorites" on your desktop computer with Eudora's browser on your PDA.

Exploring the PDA's Capabilities

The rest of this chapter explores the PDA's capabilities as sets of described solutions. As the power of the PDA increases and wireless options expand, many questions arise regarding how to best exploit these new capacities. Wireless capabilities, common concerns, and problem solving are emphasized.

E-Mail Options

The various PDA wireless e-mail options and technologies will be discussed in this section, as well as current limitations and what may be in the works to fix them.

HOW DO I VIEW ALL E-MAIL FROM VARIOUS ACCOUNTS ON MY PDA?

E-mail messages arrive to you through a number of different services and protocols. Your desktop computer stores enough protocols and client delivery types to accommodate all the e-mail you're likely to have. But the Palm OS and wireless Internet access technologies do not necessarily accommodate all e-mail delivery systems. For example, you can use Webmail, which includes Yahoo! Mail and Palm Mail. Then there is ISP-based mail with proprietary protocols, such as AOLMail and MSN. There are also corporate e-mail accounts and desktop clients, such as Outlook and Eudora.

Most PDA-based e-mail applications will not give you access to all e-mail types. Enter Gopher King, which refers to itself as an information service that delivers your messages and the information *you* need to *you* anytime, anywhere, through any device. Gopher King is one of the very few applications that allows you to view all e-mail types on a wireless PDA. Also useful is MultiMail, which ships with most new PDAs.

WHAT IF I NEED TO DOWNLOAD E-MAIL ATTACHMENTS?

Currently, the easiest e-mail application for e-mailing file attachments from your PDA is MultiMail Pro. It allows you to e-mail many file types, up to 2MB. However, the text in the e-mail itself cannot exceed 60,000 characters. MultiMail Pro is one of the most popular and powerful PDA e-mail programs. This popular program is available from download sites such as `www.zdnet.com/downloads/pilotsoftware` and `www.palmpilotware.com`.

HOW DO I CHECK MY MAIL WHEN ALL I HAVE IS AOL?

You can purchase OmniSky for AOL, available when you purchase one of OmniSky's wireless modems. The package includes the popular Minstrel Wireless Modem series. Also, the Palm Mobile Connectivity Kit, a product that works on all Palm OS–run handhelds, lets you access AOL mail. Check out HandMail 2.0 from Smartcode Software, which enables you to check AOL mail from a Palm OS–based organizer, and PocketFlash from Powermedia, although it requires wireless connection through OmniSky. Again, on the higher end, you could use Gopher King, a rather all-encompassing e-mail and information service that will let you access AOL mail from any wireless Internet-connected PDA. Interestingly, MultiMail, the mother of all Palm OS e-mail conduits, does not support AOL mail access.

HOW CAN I BE AUTOMATICALLY NOTIFIED OF E-MAIL?

Currently, PDAs do not provide always-on notification of news, e-mail, or other events the way advanced pagers do. If you want to access your mail or receive information from the Web, you'll have to log on and check. However, Motient, a Reston, Virginia-based company that provides wireless service for RIM BlackBerry pagers, will provide a pager modem and service for the Palm V. If such technology initiatives take off, PDAs can enter a huge business market where always-on instant e-mail notification is required. The prospect of not being required to check for e-mail would be immensely attractive to many customers.

Wireless Access and Data Sharing

Properly configured and equipped PDAs can wirelessly share and transmit data. Wireless access includes local devices of many types, LAN and WAN access, and teleconference management. We'll discuss what those options are, as well as current limitations and fixes, in the upcoming section, "Wireless PDA Professional Solutions."

HOW CAN I SURF WITH MY PALM VIIx WHEN I'M OUT OF RANGE FROM MOBITEX?

The only Palm modem that currently provides Internet connectivity out of the box is the Palm VIIx — through its Palm.net service via an arrangement with BellSouth's wireless data network, called Mobitex. When you are out of the Mobitex coverage area, you cannot access the Web through your Palm VIIx. However, you can purchase a cable to connect your PDA to a data-enabled cellphone, if you have one. Then, if you have a standard ISP that is reachable from your current location, you can dial it through your cellphone and access the Web that way.

HOW CAN I SHARE DATA WITH ANOTHER NEARBY PDA?

Try transferring data through Infrared ports. All newer PDAs have them, and the PDAs come with software for transferring data or even applications from one PDA to another. Third-party software is also included that will set up this connection. Check out SyncTalk FX at the download sites mentioned in the earlier section, "How Do I Find PDA Applications?"

HOW DO I WIRELESSLY CONNECT MY PDA WITH OTHER NEARBY DEVICES?

You can use one of two technologies to wirelessly access a local device, such as a printer, HotSync cradle, or network portal. First, you can use a wireless network interface card, such as the Xircom Wireless LAN Module (see Figure 2-9) for the Palm OS. The Xircom uses the 802.11b WLAN specification for high-speed wireless access. The Handera 330 PDA and Handspring Visors have CF (compact flash) cards that provide wireless 802.11b connectivity. However, the device that you're connecting to will require 802.11b access as well.

The other wireless device connection solution is Bluetooth. Bluetooth is a specification that will wirelessly connect multiple devices within a few meters of each other. No hub is needed. All Bluetooth-enabled devices within range will share data and connections. A Bluetooth-enabled PDA can thus print to a Bluetooth-enabled printer and HotSync to an enabled synchronization cradle. The beauty of the long-promised Bluetooth specification lies in the ease with which multiple devices can all share connections just by being in range. Many companies are producing Bluetooth-enabled devices for the Palm OS. For example, Pico Communications, Bluefish Wireless, and TDK Systems have each developed Bluetooth-enabled wireless access points for sharing data between Palm OS devices and a network.

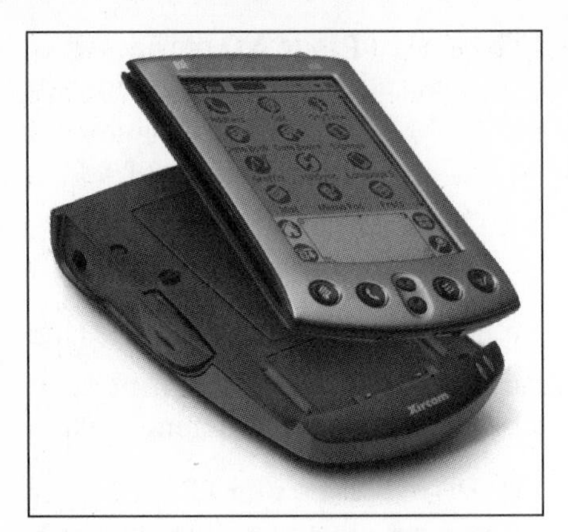

Figure 2-9: The Xircom Wireless LAN Module for the Palm PDA (photo courtesy of Xircom, Inc.)

HOW DO I TELECONFERENCE WITH MY PDA?

Evoke Mobile Web Conferencing lets you manage a teleconference with Web support from your PDA. Share visual data over a Web connection and then invite users to participate in a teleconference, all from your wireless PDA connection.

Hardware and Multimedia

In this section, some common hardware issues and questions about multimedia access are addressed.

HOW CAN I ACCOMMODATE MY PDA'S USB CONNECTION IF I HAVE ONLY A SERIAL PORT?

If your PDA comes with a USB HotSync connection (most of them do) and your computer does not support USB, you'll have to purchase a serial port connection kit and cable. This can be an issue in particular for Windows NT 4 desktop computer users, since NT 4 does not support USB. Connection kits and cabling are readily available at Palm supply sites all over the Web and at computer hardware outlets.

HOW CAN I BE ALERTED SILENTLY OF AN APPOINTMENT OR EVENT?

On the newer PDAs, you can use the silent or vibrating alarm.

HOW CAN I TALK ON THE PHONE AND SIMULTANEOUSLY USE MY PDA ON MY PDA/PHONE?

PDA/phones all have speakerphones to accommodate this requirement. You'll not have to go out of your way to accommodate this requirement.

HOW CAN I TALK ON THE PHONE AND ACCESS DATA AT THE SAME TIME ON MY PDA/PHONE?

Currently, simultaneous voice and data calls are not possible on the same wireless telephonic device. Future models from Handspring, such as those based on the Treo, may incorporate this capability. However, especially look for future PDA/Phone hybrid devices such as those based on the Symbian OS, which is discussed in Chapter 4.

HOW CAN I ADD MORE MEMORY TO MY PDA FOR ADDED APPLICATIONS?

PDAs with expansion slots to accommodate more memory requirements are becoming more and more common. You can usually purchase either 16MB or 64MB CF cards for many newer devices. Also, if applications or stored files are piling up fast on your PDA, consider offloading some data to your desktop computer.

HOW CAN I PLAY MP3S FROM MY PDA?

The Sony CLIE series supports MP3 playback right out of the box. Handspring markets MP3 players for their expansion port-enabled Visor models.

HOW CAN I VIEW MOVIES FROM MY PDA?

Many programs, such as Album to Go and FireViewer, let you view images on your PDA. However, MGI's PhotoSuite Mobile Edition (which ships with most Palm handhelds) will let you view movies (see Figure 2-10).

Figure 2-10: Screenshot of a movie displayed in a handheld with gMovie (photo courtesy of Generic Media, Inc.)

Writing, Printing, and Faxing

PDA users often complain that there's got to be an easier way to get information into the device. Wireless PDA technology makes it easier to access and download documents of all types. But once you have them, what can be done with them? Today, they can be printed and faxed from almost any PDA application, and even read at leisure. This section discusses the various options for creating, editing, and outputting documents from your PDA. Typing, writing, and drawing options are also discussed.

CAN I WRITE AND DRAW WITH MY PDA, RATHER THAN USE GRAFFITI?

If you aren't ambitious enough to learn the Palm OS shorthand Graffiti, or if, at times, you simply want to make a quick handwritten note right on your PDA scratchpad, you're in luck. Most newer PDAs provide a feature called Notepad, which lets you write anything you like in the scratchpad area, rather than evoking Graffiti.

Certain programs can provide advanced writing and drawing tools for your PDA. For example, with BugMe!, you can take screenshots of your notes and drawings and then send your notes and drawings in an e-mail via your wireless PDA connection (see Figure 2-11). Also, look into Motorola's ISketch and Cyclos' Hi-Note. All these programs expand the PDA writing or drawing area to include the entire screen.

Figure 2-11: Notetaking and drawing with BugMe!

CAN I ATTACH A KEYBOARD TO MY PDA FOR FASTER TYPING?

Yes, you can. Portable keyboards that attach through your PDA's HotSync connector are readily available. The most popular is the Targus Stowaway Portable Keyboard (see Figure 2-12). When not in use, it will fold into a size no larger than the PDA itself. It unfolds into a size just smaller than the average laptop keyboard. Palm markets a mobile keyboard as well.

Figure 2-12: The Targus Stowaway Portable Keyboard (photo courtesy of Targus, Inc.)

HOW CAN I PRINT FROM MY PDA?

For portable printing, check out the Pentax PocketJet printer, which uses an Infrared (IR) port. The size and port usage make it a very PDA-friendly choice. You can also use any other printer that can receive data via IR. However, you'll need software. If your PDA does not ship with a printing application (and few of them do), download the InStep Group's Print. The InStep Group makes several other document-oriented PDA applications, such as a program for faxing. You can also try TealPrint from TealPoint Software, which lets you print via your HotSync port (which means you can print through any USB or serial port-enabled printer), as well as Bachman Software's Bachman Print Manager. Note that these programs all let you print from most PDA applications.

HOW CAN I FAX MY DOCUMENTS WITH MY PDA?

InStep Fax, an add-on to the InStep Print program, is by far the most popular PDA faxing software. InStep Fax will fax via a mobile phone IR connection or any other connection device capable of sending Class 1 or Class 2 facsimile data. CDPD modems, such as the Minstrel series, would work just fine. Also, check out Mark/Space Softwork's Fax program, which supports many wireless device types and can transmit via IR.

Personal PDA Conveniences

This section covers features and software options that make your PDA into a personal tool. You can track your finances and your stocks and read e-books at leisure. Also discussed is choosing a PDA or accessory that provides auto-location.

HOW CAN I TRACK MY EXPENSES WITH MY PDA?

Most PDAs ship with the Expense 1 program, a basic application for expense accounting. You may also want to look up MobiTech System's Time and Expense Tracker or Solid Rock Software Group's BudgetMaster.

HOW CAN I VIEW AND TRACK STOCKS WITH MY PDA?

TinyStocks' Stock Manager manages your stock portfolio on your PDA. You can wirelessly check stock prices and synchronize the new prices on your PDA. Or, you can wait until you HotSync to synchronize stock prices. Also check out Olotek's OLOSTock, a PDA application that provides stock quotes and charts to mobile PDAs. You can easily track stocks and receive stock quotes and charts for any U.S.- or London-traded equities. Set up your own stock "watch list" or portfolio and then create charts reflecting various intervals. For example, you can make weekly, monthly, quarterly, or yearly stock charts.

HOW CAN I READ BOOKS ON MY PDA?

One of the great advantages of a PDA is that it allows you to store and view novels. Hundreds of novels are available. PDA reader applications are designed to improve your reading experience, with features that make it easy to sit back and read novel-sized documents. Good document reader programs come with a number of selectable font sizes for reducing eyestrain, and a variety of document-scrolling methods. Variable line-spacing options are helpful as well. For hands-free reading, auto-scrolling is a must. Good PDA readers include Bill Clagett's CSpot Run, TealPoint Software's TealDoc, and Palm's own Palm Reader, which features a huge number of e-novels due to its purchase of the venerable peanutpress.com books (see Figure 2-13).

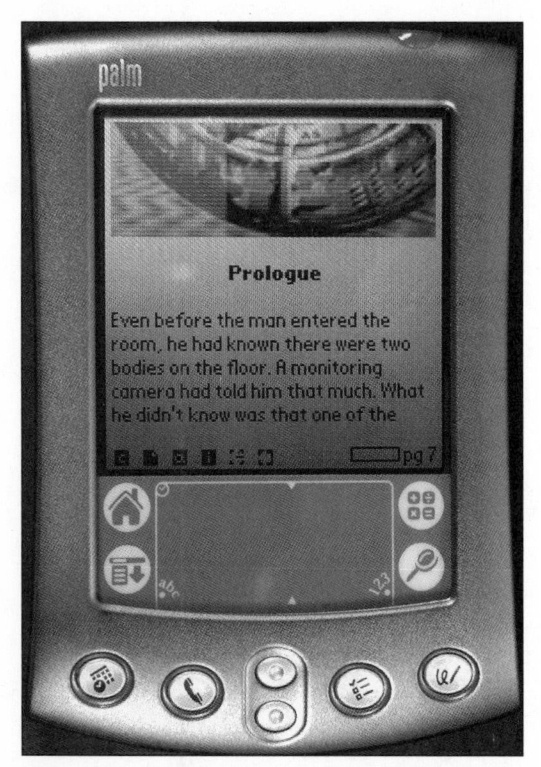

Figure 2-13: A Peanutpress novel in a PDA display (photo courtesy of Palm Digital Media)

CAN I GET A PDA WITH AUTO-LOCATION?

When you search for a location using the Palm VIIx, it also knows your location. Direct the Palm VIIx to locate "the nearest" merchant of a particular sort, and it will do just that. It will even tell you where you are on a map. Handspring makes a Springboard GPS (Global Positioning System) locator that tracks your current position via GPS satellite (see Figure 2-14). A Handspring Visor enabled this way will be able to map specified locations relative to your own.

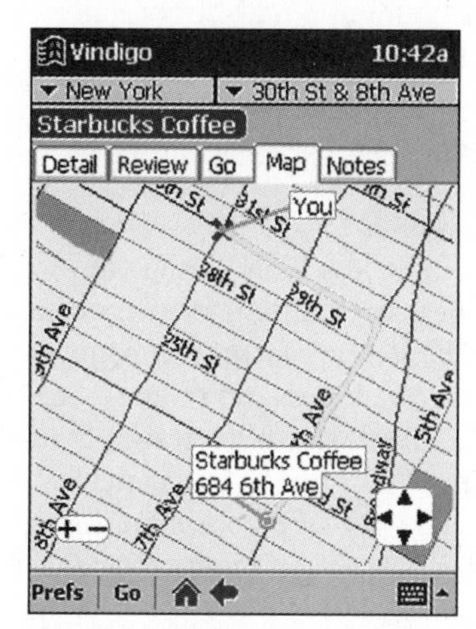

Figure 2-14: Using auto-location to find out where you are

Wireless PDA Professional Solutions

Properly configured and equipped PDAs can become essential tools for enterprises, large and small. Not simply a method for keeping road warriors in touch with the office, PDAs are easily configured mobile data tools that can work to everybody's advantage. This section discusses the PDA as a business solution.

AS A BUSY PROFESSIONAL, HOW CAN I OBTAIN AN ALL-IN-ONE PDA SOFTWARE SOLUTION?

A tremendous amount of software exists for Palm OS devices, some of it good and some not. Unlike PCs, PDAs don't provide unlimited storage space for evaluating applications that may or may not be helpful. PDA users tend to be practical, choosing applications that are of genuine utility in their occupations. All else considered, a "software suite" approach is advantageous. Why not group the best Palm applications according to occupation or vocation?

Enter Handango, which calls itself the world's leading wireless software publisher. It licenses the best applications, wraps them in a suite-like interface, and markets these suites to individuals or groups based on profession. For example, the Student Suite provides, among other features, an assignment tracker, courseware manager, and dictionary. The Teacher Suite provides a grade book, attendance manager, and calculator. There are three suites for medical professionals: one for medical students, one for residents, and another for physicians. Handango does not market "shovelware"; rather, it selects the best applications from developers such as MK Software, Ansyr Technology, FirePad, and iBrite. With screen real estate and storage space so scarce on a PDA, it's amazingly helpful to have someone else do the shopping for you.

HOW CAN I PROVIDE CUSTOMERS WITH ACCESS TO THEIR FINANCIAL DATA?

Financial institutions can build quite an advantage by giving customers access to their financial data via a wireless device. Customers would love the convenience of not only online banking but also wireless online banking; for example, you could check your account and make transactions with your PDA. Enter Aether, a mobile wireless solutions provider that develops business applications of all types. For example, Aether Mobile Finance is a complete mobile solution that enables customers of financial institutions to manage their finances anytime, anywhere.

HOW CAN I HAVE INTERNET AND WAN ACCESS ON THE ROAD?

Sierra Wireless Network Cards, such as the AirCard 300, provide a wireless WAN connection to your central network, just as if you were using your networked desktop computer back at the office. You don't have to dial up. Your PDA is instantly connected. Whatever information you can access on the network via your desktop computer — such as e-mail, databases, and so on — can potentially be accessed on your PDA. However, the connection will generally be much slower than the 19.2 Kbps capabilities of the network card. The card must run over a pricey but somewhat slow CDPD service, which is available in most urban areas in the U.S. As CDPD access improves over time, your network card's performance will also improve. The connection tends to be adequate for checking e-mail on the road and for transferring very small files.

The faster solution for urban areas where CDMA2000 and other CDMA technology is available is the Sierra Wireless AirCard 550-555. With theoretical speeds up to 153.6 Kbps, you'll be accessing data from your PDA faster than you ever dreamed possible — when the 3G technology comes of age. Using this card, you'll be able to connect to Internet, LAN/Intranet, and remote software applications at a good speed.

HOW CAN I TRACK AND DISPATCH A MOBILE FLEET?

Wireless network-enabled PDAs equipped with GPS units can aid a company with a mobile fleet of trucks or cars, helping them stay in constant contact with a dispatch center. Unit location can always be available, and employees can report on job status and receive the next work order via wireless PDA. Aesther is one company that provides such a solution, called Smart Mobile. Also check out Go-Dispatch Wirelessly Enabled PDA solutions at OnTheGo Business Solutions.

Document Editing and Messaging

A document on a PDA does not have to sit passively. Microsoft Office and WordPerfect documents can be edited on your PDA. Also, instant messaging, long a staple in the desktop Web environment, is readily available for PDAs with Internet access. The following sections suggest ways to implement these capabilities.

HOW CAN I EDIT MY OFFICE FILES BUT NOT LOSE FORMATTING?

Check out Data Viz Documents to Go for editing Word, Excel, WordPerfect, and Appleworks documents on your PDA. While editing on your PDA, the formatting is stripped. But when uploaded again to your desktop computer, the formatting is added. On your desktop computer, just drag the file you want to edit onto the Documents to Go application window. The next time you HotSync, the file will be transferred to the PDA. This system allows you to edit the content of your office documents on your PDA without altering their look and feel.

While Documents to Go is included free with many PDAs, it is not your only software choice. Cutting Edge QuickOffice performs almost the same function while adding multiunit support for synchronizing many documents across a network.

HOW CAN I SEND INSTANT MESSAGES AND EXPANDED MESSAGES?

You can quickly obtain instant message capabilities by signing up with Yahoo Messenger. Just log onto Yahoo's site with your PDA wireless connection and choose a user name and password. One of the services you'll have access to is Yahoo Messaging. MSN users, AOL members, or AOL Instant Message users can also send Instant Messages. Other services, such as ikimbo.com and airtravelcenter.com, provide Instant Message services. Of course, virtually all wireless-enabled PDAs can send SMS messages to other mobile device users. SMS messages are text entries of 160 characters or less sent over a Short Message Service server. Most PDA wireless connection services provide SMS as well.

Synchronization Options

PDA synchronization has come a long way. Today, PDA users exchange data from many applications with their desktop computer. All these shared documents must be synchronized so that both versions are up-to-date. Additionally, companies with many PDAs deployed frequently need to bring them all up-to-date with software patches, OS upgrades, and document synchronization. Several solutions are available for both requirements.

HOW CAN I SYNCHRONIZE MANY APPLICATIONS WITH MY DESKTOP?

Programs, such as PocketMirror, that come with most PDAs let you synchronize your PDA with Outlook and other familiar scheduling applications, but they do not coordinate synchronization with multiple units or work with more complex groupware applications. Pumatech's Intellisync is considered by many to be the most thorough synchronization software, allowing you to synchronize calendars, contacts, schedules, e-mail, and other applications with your desktop computer. Intellisync works with Microsoft Outlook, Schedule+, Lotus Notes and Organizer, Novell GroupWise, and ACT PIM. Also, check out Natura Software's Bonzai, and DataViz Desktop to Go.

HOW CAN I SYNCHRONIZE THE DATA ON ALL MY COMPANY'S PDAS?

Look into the AvantGo Business Server, which provides fast and easy multidevice synchronization. The AvantGo Business Server is probably the most popular method used for delivering applications and data from back-end systems to mobile devices. Another advanced software solution is Callisto Software's Orbiter System Management Server Connector 4.0, which provides central server management of multiple PDAs. Orbiter links Palm OS devices to Microsoft Systems Management Server for configuration management. IT managers can take a snapshot of software running on all PDAs and determine which need synchronization.

For a hardware solution, Palm's Ethernet Cradle synchronizes multiple PDAs to a network but is currently only available for the Palm III series. Also, look into Symbol's Four-Slot Serial port Cradle for simultaneously HotSyncing four Palm OS devices.

Creating Enterprise-Level PDA Applications

Creating specialized PDA applications is within reach of both small and large companies. Various data types and access configurations can be built right into an application created exclusively for a company's specific needs.

HOW CAN I DEVELOP A CUSTOMIZED PALM OS APPLICATION?

Creating an application to be deployed on all wireless devices in a company is quite doable. PDA-based wireless applications have to be small. The interfaces must be uncluttered, since screen space is at a premium. Quite frequently, businesses contract with PDA application developers to create customized applications for their mobile employees and find that these can be created and deployed quite quickly. One company that writes custom Palm OS applications is AppForge. Applications are written in Visual Basic 6 and are cross-platform, working on many wireless device types. A typical application includes PDA-deployed bar code scanners that record price changes in a central database.

HOW CAN I CREATE A BUSINESS APPLICATION THAT CAN SHARE INFORMATION BETWEEN VARIOUS MOBILE DEVICES?

Applications can be developed for mobile employees who need to share information across a variety of devices and operating systems. A company's mobile employees may not all be using the Palm OS. Additionally, they may be required to transmit data wirelessly to desktop computers using a variety of operating systems. Java would be an ideal tool for creating an open portal for wirelessly sharing data across many device types. Enter Wapaka, a WML and XHTML microbrowser entirely written in J2ME, J2SE, and PersonalJava. Wapaka creates Web-centric professional applications that, once written, can be deployed anywhere. These include Palm OS, EPOC and Symbian devices, as well as PCs.

After one of Wapaka's customized applications is deployed on your devices, mobile employees can create personal portals to customize their view of the application. Files can be shared via the portal using a simple drag-and-drop procedure.

ThinAirApps is another company that provides a similar service. ThinAirApps works with businesses that want to give mobile employees wireless access to corporate data. With an emphasis on security, ThinAirApps creates portals that can run on a number of wireless device operating systems. Palm OS devices, EPOC, Windows CE, and RIM BlackBerries can all have access to the same firewall-protected data while on the road. This solution is ideal for a company that does not want to incur the expense of purchasing new uniform devices for all mobile employees but would rather work with the multiple wireless devices already in place.

Summary

We've explored PDAs, their wireless potential, applications of all types, and how to accommodate a business requirement by deploying PDAs to mobile employees. In the next chapter, we move on to Pocket PCs.

Chapter 3

The Pocket PC

In This Chapter

The following topics are covered in this chapter:

- ✓ Pocket PC industry dominance
- ✓ The Pocket PC OS
- ✓ Pocket PC product lines
- ✓ Handheld PCs
- ✓ Pocket PC wireless considerations
- ✓ Pocket PC wireless devices
- ✓ Pocket PC connectivity options
- ✓ Wireless Pocket PCs in the enterprise
- ✓ Wireless Pocket PC deployment considerations

Pocket PC versus Palm

In Chapter 2, you learned that the PDA excels as a contact and e-mail management tool, mini-Web and network portal, and document viewer. In contrast, the capabilities of a Pocket PC are much closer to those of a laptop (although a Pocket PC is intrinsically more "wireless-ready"). Many types of wireless device options are available for Pocket PCs, and their portability and power has led to enthusiastic endorsements as enterprise tools.

Today, Pocket PCs are sold with terminal services and VPN client capability that has been preinstalled, making it quite possible to run applications remotely, access a server, and surf "whole" Web pages. Also, e-mail applications and clients for the Pocket PC will look very similar to your desktop counterparts. All in all, you will feel as if you have simply taken your desktop computer on the road.

Pocket PCs have relatively large and clear full-color screens. They can store much more data than a Palm and can easily accommodate expansion modules of many types. The Pocket PC OS has the familiar look of Windows. Thus, deploying Pocket PCs to a number of mobile employees does not require a trek into unfamiliar territory. While an enthusiastic community of Palm application developers has created literally thousands of PDA programs, Pocket PCs use Windows programs — the same applications that everyone knows. For someone who is trying to learn minicomputing for the first time, this familiarity is a huge advantage.

OS-Driven Advances

In recent years, Pocket PC manufacturers have hit the stage with a vengeance. Microsoft's Pocket PC partners are the most familiar names in computing (for example, Hewlett-Packard and Compaq). In a virtual alliance with Microsoft, Pocket PC models have been released in lockstep with Microsoft's new Pocket OS, each release generating the familiar Microsoft media hoopla. Simultaneously, companies such as Compaq, Hewlett Packard, Casio, and Toshiba have provided more memory, better color, and more expandability with each new device. They've created a powerful portable computer that not only lets you access data while on the road but also allows you to run applications remotely.

One common goal is the "Netscape-ization" of the Palm, to gradually relegate the PDA giant to a sliver of its current market share. Microsoft and its partners do this by sprouting new features every few months. Since it first introduced the Windows CE operating system for handhelds in 1997, Microsoft has developed significant improvements with each release. For example, the current Pocket PC OS, Pocket PC 2002, is a Windows XP-like powerhouse. Looking at this latest release, you get the feeling that not much more could be squeezed into a Pocket PC.

Microsoft and its Pocket PC partners view wireless as "the future," and they mean business. Together, they have deep pockets and can weather early missteps and failed expectations by simply marketing a new model. It's no wonder that Palm may feel besieged.

Pocket PC as Enterprise Solution

To accommodate enterprise application development in a big way, Microsoft has provided and actively promoted a robust developer's toolkit. With this toolkit, application developers familiar with Visual Basic or ActiveX data objects will have no trouble creating usable and highly appreciated tools for the Pocket PC environment. Large and small businesses are just now getting a taste of Pocket PC's capabilities. For example, a programmer can create a form with hundreds of fields. Mobile employees can click through choices with a touch-screen stylus and fill out a field-data form in a few minutes, their responses automatically saved in a database server miles away. Scary.

Beating Planned Obsolescence

Together with Chapters 1 and 2, this chapter helps individuals and corporations choose the best wireless communication device for the job. Planned obsolescence and the shifting fortunes of wireless device manufacturers make such decisions crucial. Whether you purchase a single handheld device or an entire fleet, you should expect to find software for (and be able to easily upgrade) your purchase for years to come rather than purchase equipment that becomes obsolete in less than a year. That's how it works with PCs, right? Not so fast. Pocket PC wireless communication development happens at an even faster pace than what you've come to grudgingly accept from our yearly desktop computer upgrades. So, if you are buying a device for yourself or your company, let the buyer beware.

Today, wireless handheld devices are used in all phases of a business: data collection, secure radio access, mobile portal to corporate Intranets, recording sales information, mapping, and wireless access to enterprise applications of all types. Developing and then deploying a highly customized wireless application is not cheap. Creating a solution to fit a Pocket PC that will have minimal manufacturer support within a year could be devastating for a small or mid-sized firm.

In this chapter, you learn about the different types of Pocket PC and handheld computers, their best uses, and their wireless options. We'll discuss how various large and small corporations, research facilities, and civic institutions have effectively deployed wireless Pocket PCs. Whether you are an individual or a corporate communications officer seeking a wireless business solution, wireless options exist for you.

Pocket PC Background

In terms of operability and use, there are three types of Pocket PCs: original Pocket PCs with the Windows CE OS, Pocket PCs running on Pocket PC 2000 or 2002 OS, and handheld computers.

The "original" Pocket PCs, deployed on the Windows CE OS versions 1 through 3, largely represent Microsoft's period of missteps in the handheld market. Comparisons to early versions of Windows are apt. During this period, the Palm PDA enjoyed complete dominance of the handheld computer market.

The Pocket PC 2000 Device

Microsoft weathered its mistakes and, in mid-2000, introduced the Pocket PC 2000 OS. This release was immediately deployed on devices whose successors could ultimately become as ubiquitous as the PC. Pocket 2000–powered Pocket PCs had monochrome or color screens measuring 3.5 to 3.8 inches diagonally. They had 16MB flash memory for storing and running the operating system and core applications. The units also provided 16 to 32MB RAM for adding more applications, as well as file storage. Some models were expandable via a CompactFlash (CF) card port. (See the sidebar titled, "Pocket PC Expansion Cards.")

Many Pocket PCs running with Pocket PC 2000 are powered by 131 or 150 MHz SH3 chips. Current models use the Intel 206 MHz SA-1110 StrongArm RISC processor. The change in processor makes software old-to-new compatibility difficult and OS upgrade impossible, though the older processors feel anything but slow. Nonetheless, Windows Pocket PC 2002 is so feature-rich that the stronger processor becomes a necessity, as is always the case with Windows upgrades.

Color displays on "original" Pocket PC 2000 devices are not all that bad. Screens showing 65,000 colors are common, and TFT reflective displays that improve visibility in indoor and outdoor lighting are available on some of these models. In fact, Compaq seems to have hit a very workable "sweet spot" with its 12-bit 4,096-color display. Rather than deploy processing power and battery life to deliver 65,000 colors (which is really only required for multimedia applications), Compaq provides an exceptionally bright screen that is genuinely readable both indoors and out. All other factors being equal, a Pocket PC with a 4,096-color display will have noticeably longer battery life.

The Pocket PC 2002 Device

With the release of the Pocket PC 2002 operating system in late 2001 (at media events staged simultaneously in San Francisco and London), Microsoft introduced the current generation of Pocket PCs. Appearing with Microsoft representatives were spokespeople for almost every Pocket PC manufacturer, each of whom unveiled devices that would put the new OS to good use. Within two months, almost every commercially available Pocket PC was running with Pocket PC 2002. In staging such a broad tie-in, Microsoft set a new standard of minimum hardware requirements for the new OS. These requirements ensured a high degree of interdevice compatibility, making it virtually impossible for consumers to buy a really, really bad Pocket PC 2002.

Pocket PC Expansion Cards

CompactFlash (CF) cards are generally used for I/O and connectivity expansion, such as wireless LAN card, keyboard, modem, camera, and increasing memory and data storage. Multimedia CF devices, such as digital cameras and MP3 players, are also available for Pocket PC 2000 models. Some Pocket PC CF ports can accommodate Type I and Type II cards, with a larger selection of devices available as Type II. The Type II–capable CF port opens a window to a large variety of wireless connectivity devices. In fact, many newer Pocket PCs using the Pocket PC 2002 OS do not accommodate Type II CompactFlash cards. There have been frequent complaints that you must purchase expensive sleeves and adapters before expansion is possible.

Another increasingly favored expansion media is the *Multimedia Memory Card/Secure Digital card*, known as *MMC/SD*, or sometimes just *SD card*. Popular because of its postage-stamp size, many devices are being developed for the SD card format. Currently, multimedia options and memory storage are the only devices available for the MMC/SD expansion slot, although many others are on the way. For example, 1GB SD cards and a Bluetooth radio are currently in development. Pocket PCs without MMC/SD expansion slots will be unable to utilize many newer accessories.

A PC card slot provides immediate expandability to the Pocket PC. PCMCIA cards common to laptops could be developed for Pocket PCs. Unfortunately, none of the "original" Pocket PC 2000 devices provide a PC card slot, and only a very few Pocket PC 2002–driven devices do.

To facilitate future upgrades, all Pocket PC 2002 devices must have at least 16MB of ROM flash memory, 32MB RAM (preferably 64MB), a 65,000-color display measuring at least 3.5 inches diagonally, and at least eight hours of battery life. The Pocket PC 2002 is a multimedia OS, so units must provide stereo sound and a stereo headphone jack. External storage for MMC, SD, and CompactFlash is required. Making the postage stamp–sized MMC/SD card a standard feature ensures that manufacturers will produce a steady stream of expansion cards for all types of devices, including modems and WLAN and PAN expansion cards.

An important under-the-hood requirement for Pocket PC 2002 units is that RAM memory must be retained for at least 72 hours in the event of a low-battery unit shutdown. In fact, 168 hours is preferable. Such a requirement ensures that you won't lose your applications and files if you run out of battery on the road. While Microsoft's hefty hardware requirements guarantee Pocket PC product quality, the quality will come as a shock to the pocketbook. Currently, Pocket PCs outfitted with wireless LAN or Internet access cost about $1,000 dollars. Clearly, Pocket PCs are being positioned as business tools, rather than a consumer convenience à la Palm. Corporations who deploy Pocket PC 2002–powered devices as enterprise solutions can hopefully count on Microsoft and its partners to provide continued software and hardware support for business applications. Such strict requirements, however, would definitely tie the hands of manufacturers, who would also want to market a lower-cost (for example, grayscale) smaller-screen Pocket PC with 16MB of RAM.

Current Pocket PC Product Lines

The following sections explore some of the current Pocket PC product lines. Only features that make each product line particularly popular are highlighted.

The Compaq H3800 Series

The Compaq H3800 series (see Figure 3-1) has a popular ultrabright 3.8-inch TFT active matrix display, as well as voice control. You can dictate commands to Calendar, Contacts, and other core applications. At less than 3.5 inches wide, this powerful unit still fits in a shirt pocket. Newer Compaq Pocket PCs provide at least 64MB RAM and 10-hour battery life. To add modules, you must *first* purchase expansion packs from Compaq specifically designed for your model of Pocket PC.

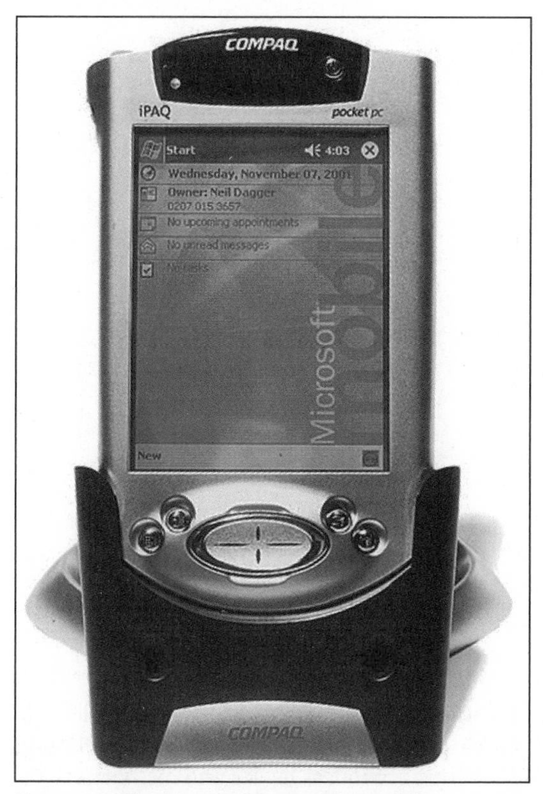

Figure 3-1: The Compaq H3870 Pocket PC (photo courtesy of Compaq Computer Corporation)

The Hewlett Packard Jornada 560 Series

The Hewlett Packard Jornada 560 series (see Figure 3-2) includes 64MB SDRAM and editable Flash ROM for storing backups of important files. Files saved in ROM will not be affected by low-battery shutdowns. At only 3.1 inches wide, this HP Pocket PC line is known for its long battery life, a confirmable 14 hours. A Type 1 CompactFlash slot is provided. The Jornada line is known not only for its longer battery life but also for its large and comprehensive software selection, both in the unit's core applications and in supplementary applications on the accompanying CD. Note, however, the omission of both a Type II expansion slot and MMC/SD slot, which makes outfitting the Jornada with future wireless devices problematic.

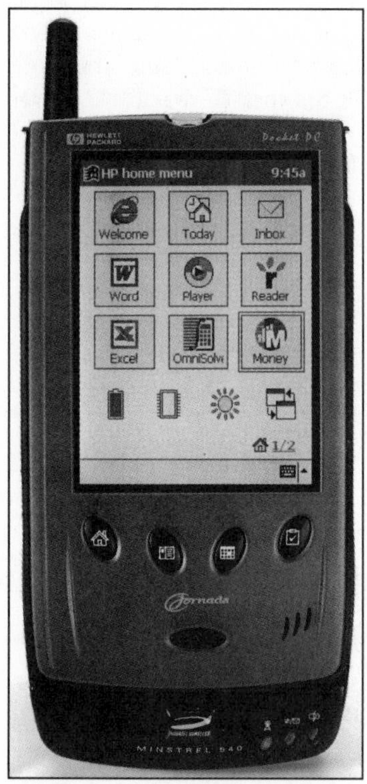

Figure 3-2: The Hewlett Packard Jornada 560 Pocket PC
(photo courtesy of Hewlett-Packard Company)

The Casio E-200 Series

The Casio E-200 series (see Figure 3-3) is popular for its advanced multimedia features. The Casio E-200 is unique in that both CompactFlash and SD cards can be used simultaneously, doubling your expansion potential. That means it is possible to utilize a wireless device and memory expansion card. Both the built-in speaker and stereo output are of genuinely high quality. Using the expansion slots as storage, the Casio Pocket PC can double as a decent portable MP3 player.

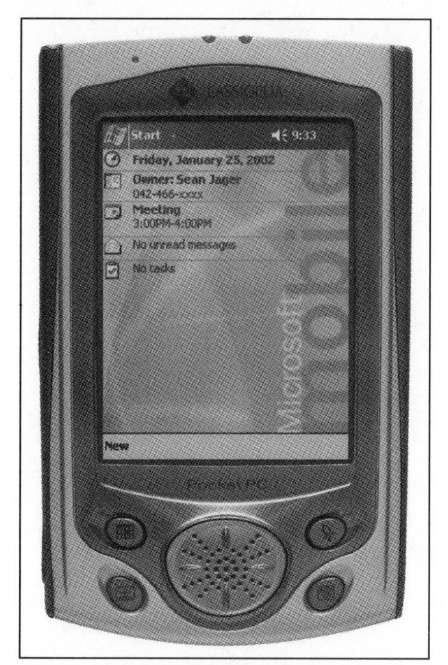

Figure 3-3: The Casio E-200 Pocket PC
(photo courtesy of Casio, Inc.)

The Toshiba e570 and URThere @migo

Another current Pocket PC that deserves mention is the Toshiba e570 series (see Figure 3-4). Toshiba supports PC cards, such as PCMCIA-style wireless LAN cards — right out of the box. Toshiba provides two built-in expansion slots, one for SD cards and one for CompactFlash Type II cards. The Toshiba comes standard, with 64MB RAM, 32MB Flash ROM, and a 3.5-inch 65,000-color TFT display. Like all Pocket PC 2002–powered devices, it has plenty of software to put all that power to work. At eight hours between charges, however, the battery life is nothing exceptional. One unique Toshiba Pocket PC feature is data interchangeability with newer Toshiba laptops. Toshiba laptops have SD slots as well, so Toshiba Pocket PCs can share files with laptops simply by exchanging SD cards.

Finally, another high-powered Pocket PC 2002 unit that supports easy expansion is URThere's @migo-600c. As well as standard Pocket PC features, the @migo 600 series has a Type II CompactFlash expansion slot. Like other Type II–friendly units, the @migo opens the door to the widest array of wireless communication cards. You'll have no trouble finding a wireless modem, WLAN, or network card to work with your device.

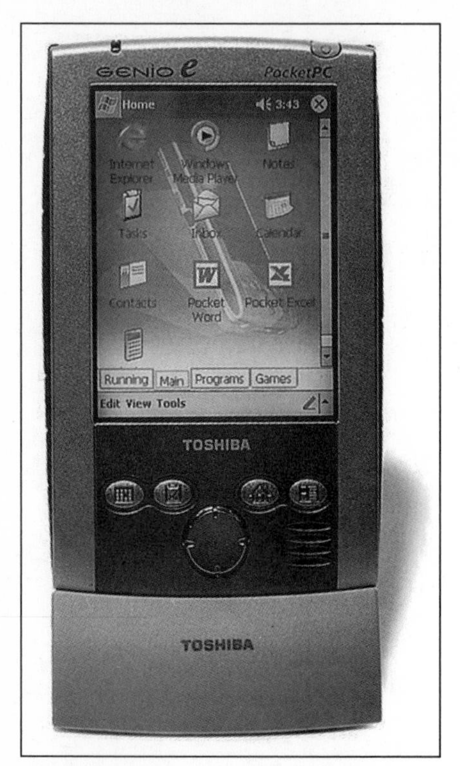

Figure 3-4: The Toshiba e570 Pocket PC (photo courtesy of Toshiba Corporation)

Handheld PCs

The handheld PC is a class of portable computer not well known to consumers. Because of its price — hovering just under a thousand dollars — you may be tempted to ask, "Why not just buy a laptop?" The fact is, the handheld PC comes closer than any other computing unit to having a "wireless nature." It is positioned to become an integral part of any potentially mobile workforce. Any enterprise that sets a handheld PC to task will quickly wonder how they ever did without it.

Those who have seen the handheld PC may recognize it as one of those checkbook-sized mobile computers with the mini-keyboard (see Figure 3-5). The nearby sidebar, "How to Identify a Handheld PC," describes the most unique features of the Handheld PC.

How to Identify a Handheld PC

Handheld PCs such as the HP Jornada Handheld series and the NEC MobilePro are unique in form and features. These units are capable of displaying half of a VGA screen, as opposed to Pocket PCs, which display a quarter. Measured key-per-key, the standard handheld PC keyboard is around 90 percent the scale of a desktop keyboard. Handheld computers provide data entry with a touch-screen stylus, as well as a keyboard. Most models provide a port for VPN access.

Figure 3-5: The Hewlett Packard Jornada 720 Handheld PC (photo courtesy of Hewlett-Packard Company)

Enterprise and Wireless-Ready

Handhelds can be a "thin client" by nature. They have a terminal server client built into ROM, and you can remotely access applications running on a Windows 2000 server. Additionally, handheld PCs have instant on/off capabilities, with sufficient battery life to complete a day's work while running wirelessly. Designed to withstand a drop from 30 inches, all core applications and OS are stored in ROM, protecting important files from accidental removal. Handheld PCs browse the Web using a full-featured browser equivalent to the desktop Internet Explorer 4.

Powerful and portable as they are, handheld PCs are not flying off the store shelves. They look hefty and industrial and retro-70s – not very inviting. Too big for a shirt pocket, you can easily envision them stuffed into a briefcase or glove compartment. Sleek and sexy, they are not.

Also worth mentioning under the rubric of handheld PC are Casio's EG-800 and Symbol Technology's Symbol 2700. These units do not provide a half-screen VGA monitor, but they are both rugged or rubberized portable computers built to withstand a little stress. If you are looking for mobile workforce portability that fits in a shirt pocket, consider these devices. Marketed as enterprise-level Pocket PCs, they come off the shelf with wireless connectivity hardware bar code data capture capabilities.

Hewlett Packard Jornada Handheld PCs

The newer HP Jornada handhelds have built-in 56K modems, a 640 × 240 screen, and drivers for a number of wireless LAN adapters written into ROM. The Jornada 820 in particular has a full 640 × 480 display. The Jornada handhelds have eight to nine hours of battery life, and uniquely, two Type II card slots: one PC-card style similar to laptops, and the other a CompactFlash card slot. Jornada handhelds also include a proprietary smart card slot. Smart cards can store information about the user to ensure secure access to networked information.

NEC Handheld PCs

NEC's line of handheld PCs is aimed at enterprise users. Current models range from the MobilePro P300, which resembles a Pocket PC, with a 3.8-inch 65,000-color screen, to the MobilePro 880, which, though a handheld, supports a full-size VGA screen. The MobilePro 700 series provides drivers that support multiple wireless connection devices and remote VPN access. It fully supports thin client applications through the included Microsoft Terminal Server Client. The NEC MobilePro 790 has a 168 MHz VR4121TM 64-bit MIPS microprocessor, a built-in 56K v.90 modem, a voice recorder, and two Type II expansion slots.

The Pocket PC 2002 Software Powerhouse

This section takes a quick look at software included with Pocket PC 2002 and handheld PC units. The interplay between RAM and ROM on Pocket PCs is worth mentioning, as well as the ability to run software from an expansion card.

Even in well-equipped Pocket PCs, memory for storing systems files, applications, and other files is limited. You cannot engage in an installation free-for-all, trying on lots of programs for size all at once. Pocket PC 2002 devices come with plenty of preinstalled programs — core applications that are stored mostly in ROM. These programs are not affected by low-battery shutdown problems or other errors that can affect files stored in RAM. Some Pocket PCs set aside a portion of ROM so that you can save critical files of your own.

Embedded Applications

Certain embedded applications — such as Calendar, Contacts, Tasks, and Notes — are included on all Pocket PCs. Programs preloaded onto your Pocket PC 2002 include Pocket Outlook, Pocket Internet Explorer, Pocket Word, and Pocket Excel (strangely, no pocket version of PowerPoint exists). Driver support for a multitude of wireless devices is already installed, as well as ActiveSync, Voice Recorder, NoteTaker, MSN Messenger, Microsoft Transcriber, and Microsoft's Terminal Services Client.

PC manufacturers can also preinstall their own proprietary utilities. For example, HP Jornadas come with LandWare OmniSolve business calculator, Developer One Code Wallet Pro, and a host of Jornada utilities such as a camera application, a backup program, and a desktop application switcher.

The Compaq iPAQ will include Compaq's own iPAQ FlashROM File Store, iPAQ Task Manager, and other fun utilities. Most Pocket PC 2002 units support a micro-Java application of some sort — for example, HP Microchai. A small set of development tools is part of the standard Pocket PC 2002 installation.

On the CD

Software included on CD is optional. Installing these programs involves synchronizing with your desktop computer using the included ActiveSync application. CD software includes optional drivers and multimedia files such as MusicMatch Jukebox, Audible Player for Pocket PC, PhatWare, and HPC Notes.

Low-Memory Workaround

As mentioned earlier, Pocket PC memory is limited. If you need to run a hefty application in an environment short on available memory, there are two steps you can take:

✓ You can run an application from external media. The Pocket PC lets you install applications on expansion cards. Depending on the type of media card and application, you can bypass installing the bulk of installation files onto the Pocket PC and run the off-unit program externally.

✓ If you have access to a remote server, run the application across your network. You don't need to install a redundant application if it already exists in your enterprise. With the Pocket PC 2002 Terminal Services Client, you can share not only files but also applications.

Going Wireless with Your Pocket PC

Chapters 1 and 2 discussed hardware features — specifically, the different features available according to model type and manufacturer. However, with the exception of expansion slot and RAM differences, Pocket PCs running the current Microsoft OS pretty much conform to a uniform set of features. The applications available for Pocket PCs will be quite familiar to Windows users; thus, there's not much need to describe how they run in miniaturization. We'll move on instead to a discussion about wireless Pocket PC options.

One interesting difference with the Pocket PC 2002 OS is how Pocket PC 2002 manages applications you are trying to close. When you click the "X" in the upper-right corner of a window to close an application, the current application screen disappears, but the program does not necessarily close. Rather, the OS monitors RAM usage and will close idle applications as necessary to free up memory. However, your device can still slow noticeably before Pocket PC 2002 will begin to shut down open applications. Manually quitting an application requires a third-party program such as White Dolphin Solutions' PocketSTM and Grundle Software's FastTask Plus.

Determining Mobile Connectivity Requirements

Whether you are choosing a Pocket PC wireless communication option for yourself or on behalf of a business, the following considerations will be helpful. The least expensive way to go wireless is with a long-term service agreement, but before getting locked into a regular expense that may not be necessary or desirable, make a mental checklist of the following questions:

- ✓ **Application requirements:** Will you need to synchronize data over your connection? Synchronization examples include updating e-mail in various inboxes, or remote calendar management. Will you require access to a database that keeps track of customer transactions completed by mobile employees? Will you need to remain connected while using an application remotely? If you are going wireless for one of the above reasons, think about setting up your Pocket PC as a "thin client" via terminal services client.
- ✓ **Geographical coverage:** Does everyone in your office need to share a connection to the Internet and to a local network? If so, look into setting up a wireless LAN with access points reachable by all affected employees. Will you be making telephone calls to, as well as accessing data from, specific remote locations across the country? Set up a virtual private network using your Pocket PC's VPN software.
- ✓ **Data amount:** Will you use an air link to access e-mail only, or to wirelessly transport large files? If you require only occasional e-mail access, consider connecting to the Internet via your data-ready cellphone. You'll spend very little on equipment, and you can inquire if your cellular provider has a data-access plan that does not deplete your voice data minutes. If you expect to be doing lots of file transport, consider a wireless WAN option. Or you can park yourself as close to a large urban area as possible and invest in a wireless modem, such as that manufactured by Sierra Wireless, Enfora, or Novatel.
- ✓ **Wireless service type:** Will you or your employees require Web surfing, access to full Web pages, as well as e-mail, remote synchronization of files, and remote application access? Services such as GoAmerica and OmniSky create an environment that most closely resembles a desktop Web-surfing experience, complete with customizable Internet portal pages ("MyPage"), access to all Web sites, and the ability to download and save files from Web sites onto your Pocket PC.

Wireless Device Types

This section discusses wireless communication access options. Some of the names of device manufacturers and service providers have been mentioned in previous chapters. For example, Sierra Wireless and Novatel will be familiar from the last chapter, as well as Xircom's Wireless LAN. Since some Pocket PCs accommodate Type II cards, many manufacturers of wireless devices can develop Pocket PC modules without abandoning their existing form factor.

There are other names as well, especially in the wireless LAN arena. Because Pocket PCs are such natural tools for empowering a mobile workforce, you'll have many wireless LAN and wireless WAN card manufacturers to choose from.

CLIP-ON, COMPACTFLASH, OR SD SLOT

CompactFlash and clip-on "sled" devices are readily available for the Pocket PC. With the expansion sleeve or jacket that some Pocket PCs require to accommodate PC Cards or Type II cards, the units become bulky — no longer pocket-sized. However, many manufacturers are currently developing postage stamp–sized devices for the SD slots. These devices will not significantly increase the Pocket PC's bulk or weight.

WIRELESS WAN AND LAN MANUFACTURERS

Wireless WAN PC cards for Pocket PCs are currently available from Sierra Wireless (see Figure 3-6) and Novatel. Many players have jumped into the wireless LAN arena. For example, Cisco-Aironet, Lucent/ORiNOCO, and Symbol Technologies all make wireless LAN PC Cards.

Figure 3-6: The Sierra Wireless AirCard 500 (photo courtesy of Sierra Wireless, Inc.)

Wireless LAN CompactFlash devices are popular as well. Note, however, that not all cards will work with the newer Pocket PC 2002 OS. Makers include Sierra Wireless, Z-com, Socket Communications, and Symbol Technologies. The Proxim Harmony CompactFlash card (see Figure 3-7) is particularly popular. As mentioned earlier, most of these companies have popularized PC-card and CompactFlash devices. Soon, SD slot-sized devices will be available and will be considerably more convenient. Size alone can be an important consideration, since most Pocket PC carrying cases will not accommodate a bulky clip-on device or shield an expensive PC card or CF device from the elements.

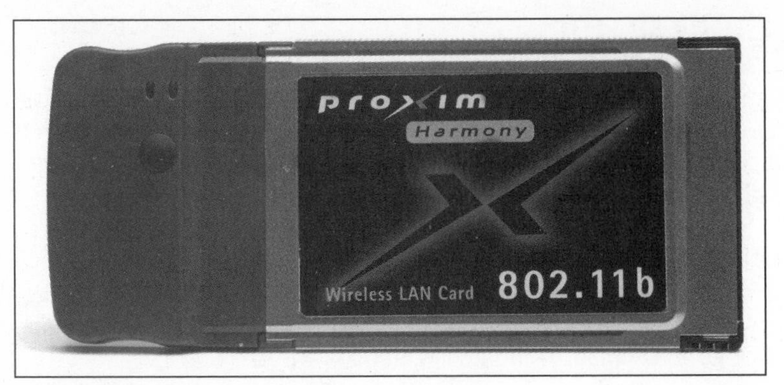

Figure 3-7: The Proxim Harmony CompactFlash Wireless LAN card (photo courtesy of Proxim, Inc.)

WIRELESS ACCESS OPTIONS

Starting from the ground up, the following sections describe some of the most effective ways to go wireless with your Pocket PC.

WIRELESS ACCESS VIA CELLPHONE You can access the Internet using a data-capable cellphone with card and cable attachments from Socket Communications and Xircom, among others. While this option can be convenient in a pinch, bandwidth will more than likely be on the order of 8 or 9 Kbps — sufficient to send and receive e-mail, but not to do much else. Also, in most instances, you'll be charged for your cellphone minutes when you use your cellphone as a modem, although some cellular plans have a special lower data-access rate. Some solutions, Option's SoftRadius for example, let you access a data-ready GSM cellphone through your Pocket PC's serial port, which can be helpful if the CompactFlash expansion slot is being used by another device. This solution allows continued access to the Internet or e-mail without having to free up the CF port.

WIRELESS MODEM OPTIONS Modems are available for Pocket PCs as detachable clip-ons or plug-in modules, such as CompactFlash and PC cards. Sierra Wireless provides popular modem cards for many Pocket PCs — available for single purchase or packaged with OmniSky, a wireless access plan. Whereas cards are usually not device-specific and can be used in most Pocket PCs that support the card's technology, clip-on modems, such as the Novatel Minstrel, are designed for specific Pocket PCs. For example, the Novatel Minstrel 540 (see Figure 3-8) clip-on wireless modem can be used with only the HP Jornada 540 Pocket PC. The Enfora PocketSpider is a CDPD modem that was originally available for Casio Cassiopeia users only. Now, as a CompactFlash unit, the PocketSpider works well in virtually any Pocket PC that has a CompactFlash slot.

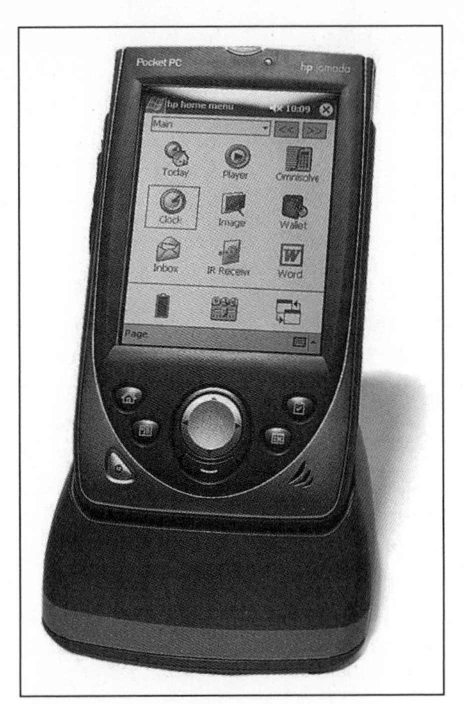

Figure 3-8: The Novatel Minstrel 540 Modem for the Jornada 540 Pocket PC (photo courtesy of Novatel Wireless, Inc.)

WIRELESS LAN OPERATION Wireless LAN cards, which allow access to a local area network (LAN), are available from many manufacturers. Currently, most wireless LANs use the 802.11b specification. Setting up a wireless LAN requires an access point that provides a physical Ethernet connection to the network, as well as a wireless transmitter via antenna to all wireless units within range. A wireless LAN provides connectivity to local drives and Internet access within 30 to 100 feet indoors. Wider access is possible with fewer walls; thus, if you are separated from the wireless access point by a number of walls, access range will be diminished. In an outdoor area with "clear site" between your wireless device and the access point, LAN access can be more than 500 feet. A wireless LAN enables access not only to a local network but also to the Internet — provided the network itself has Internet access. Wireless LAN access does not incur a service charge. Once you've paid for the equipment, all wireless-enabled devices can access the network and, with the right software, the Internet. Depending on conditions, wireless LANs provide data throughput at 2 to 11 Mbps.

TERMINAL SERVICES CLIENT ACCESS Accessing the network as a terminal services client provides Internet access as well as access to remote applications. Symbol Technology's Symbol 2700 is unique in that the Pocket PC provides a built-in link to the open architecture of Symbol's Spectrum 24 wireless LAN.

WIRELESS WAN ACCESS Wireless WAN cards and devices provide up to 19.2 Kbps Internet access over the Cellular Digital Packet Data (CDPD) air link. The CDPD network is a standard cellular network overlay and is available through carriers such as AT&T Wireless, Bell Atlantic Mobile, and GTE Wireless. Wireless communication access via CDPD varies in cost and availability throughout the United States. Before purchasing a CDPD wireless device, inquire about CDPD access with providers in your area. Popular wireless WAN devices include the Sierra Wireless AirCard series, as well as the Novatel Sage and Minstrel series.

PURCHASING WIRELESS SERVICE WITH A DEVICE Often, both individuals and business find it easier to purchase a complete wireless solution, buying both the modem, or card, and perhaps a year's worth of regular service simultaneously. Some providers offer a package that includes a Pocket PC, wireless access device, and monthly service. Wireless access over an air link usually runs in the order of $29 to $39 per month, all you can eat. The charge is not based on data transfer amount. Curiously enough, OmniSky announced in late 2001 that it would begin charging Compaq iPAQ users a premium amount, based on their observations that iPAQ users access wireless services much more than HP Jornada or other Pocket PC–brand users. The most popular service providers are OmniSky and GoAmerica. Since yearly plans are offered, purchasing a plan through OmniSky or GoAmerica can save money. These companies are service resellers and will get you connected with the cellular provider in your area that offers a CDPD overlay. By purchasing a wireless unit and Pocket PC through OmniSky or GoAmerica, you won't have to worry about compatibility hassles, since both devices are sold as a unit.

ENTERPRISE END-TO-END PROVIDERS Corporations that need to deploy a mobile wireless solution with immediate accuracy will benefit from one of many end-to-end solution providers, each with its own area of industry expertise. More specifically, these mobile solution providers develop applications for the Pocket PC, suggest hardware combinations, and provide the necessary air link and bandwidth to get your application up and running. Because their services can run into the thousands of dollars, you should approach them knowing exactly what you want to accomplish, and how much of a learning curve your employees are up for. A few of the more well-known mobile solution providers are the following:

✓ Abaco: Creates custom enterprise resource planning solutions optimized for wireless LANs and WANs. Creates plans that work well even in low-bandwidth connections.

✓ Aether: Creates complete wireless enterprise solutions. Aether will purchase mobile devices and peripherals, provision the necessary wireless service, and develop a customized application for the most advantageous deployment method for your company. If necessary, the company will also host the solution as an application service provider.

✓ Paradigm4: Specializes in applications for public safety and education. Paradigm4 is also a wireless service reseller.

✓ Ameranth: Creates secure, wireless LAN solutions especially for hospitality (hotel), healthcare, and retail industry systems. Designs integrate point-of-sale systems for completing customer transactions wirelessly.

Enterprise Wireless Solutions

This section discusses several successful and much-heralded enterprise-level Pocket PC solutions. The problem presented in each case study in the upcoming section, "Wireless Device Deployment Case Studies," is how to deploy a device with a user-friendly interface and application; the goal is to create a streamlined data entry process for mobile employees. Some situations involve gathering data in environments that may be hostile to high-tech equipment, such as the Port of Seattle, a Utah police precinct, Calgary emergency service, or the Vancouver wetlands.

Although the discussion focuses on macro-level wireless solutions, the concepts can also be applied to smaller enterprises. In fact, one common thread that runs through most of the solutions is the use of familiar programming tools. Either one of the enterprise administrators had some programming background, or a consultant was hired that applied a basic set of data objects in the Pocket PC environment. In defining the ideal application for your own mobile workforce, you'll find that once you know what you want to accomplish and what specific hardware to use, the process will move along rather quickly. After you are familiarized with the enterprise solutions presented in this section, we'll give you some points to consider when developing your own Pocket PC-based wireless solution.

Determining Need

First, determine if your business really needs remote or mobile connectivity. Do field employees frequently contact the office needing new destinations, customer status updates, or messaging? Do employees collect data at one location and then log it at another? For example, will you be accommodating offsite customers, citywide residential data collection, land surveying, or the filling out of medical forms? Do staff members frequently attend trade shows or conventions and need access to files while on the run? If the answer is yes to any of the above, then your company could benefit from a mobile wireless solution.

Determining Size Requirements

Does the mobile wireless device need to be as small as possible, or can it remain in a glove compartment or car seat? Will the mobile employee be interviewing customers while checking off form fields on a Pocket PC? If shirt-pocket size is a necessity, consider an SD card-size modem in a Pocket PC with a small form factor — the Toshiba e570, for example.

Ruggedness Requirement

Does the device need to be rugged, resistant to damage from adverse weather or dirt and dust? If so, consider a handheld PC, such as a Symbol PP2700 or Casio EG-800, rather than a Pocket PC. These are rugged devices geared for industrial use. If "instant on" is required, also consider a handheld PC. One press of a button and applications are available almost immediately.

Wireless Solution Development Tips

Here are a few basic tips for application development in the enterprise:

- ✓ **Leverage all available experience.** Pick the brains of anyone in your firm who has worked with or helped develop a custom application for the workplace. Even if the application was not a wireless Pocket PC program, such insight can still prove valuable.
- ✓ **Deploy first to a few tech-savvy employees.** You will need to work out the bugs and kinks in your wireless application. When it is nearly developed, let one of your hard-charging technically oriented employees test it while on the go. Figure out what needs to be fixed before the code is entirely put to bed and the application is completed.
- ✓ **Develop an application for a single department first.** If possible, try to avoid having everybody "go live" with the wireless application at once. Let one department live with it for a bit and then build outward.
- ✓ **Develop for one mobile device.** Communications officers or IT department heads usually get the idea for a custom mobile application when they see tech-savvy employees accessing all kinds of desktop applications from their own wireless devices. Everyone has a PDA or a Pocket PC these days. If you want to leverage existing equipment without incurring the expense of purchasing one device type for all mobile employees, write a Java application (this is probably the most cross-platform solution). Otherwise, write for one device so that you can standardize the application and create uniform teaching aids for bringing everyone up to speed. Later on, troubleshooting will be easier if everyone has one device type.

Are there people in your enterprise with Visual Basic 6 skills or familiarity with ActiveX data objects? Possessors of such skills will have an easy time developing an application with the Windows CE or Pocket PC 2002 developer's toolkit.

Wireless Device Deployment Case Studies

This section provides examples of successful wireless solutions in the enterprise.

RESEARCH DATA COLLECTION

The Motion Analysis Laboratory at Children's Hospital in Richmond, Virginia, researches the movement and motion of neuromuscular disease victims. By documenting patient gait and other motion-related observations, neurologists gain valuable insight into a variety of disease processes. The hospital's Motion Analysis Laboratory, launched in 1988, combines state-of-the-art camera equipment and other recording technology with skillful recognition of neuromuscular disease symptoms. Patient data intake involves filling out 12 forms, each with multiple sheets of paper. Obviously, a way to replace manual data entry would be most welcome. An ideal solution would be to deploy researchers with mobile wireless devices for data intake. Patient data could be recorded right onto a central database. Fortunately, one of the main research directors at Children's Hospital was familiar with Visual Basic and consequently developed a form built with database objects for the Windows CE environment. The application was developed, and mobile units were deployed to a relieved staff who marveled at the timesaving and reduced drudgery in collecting patient data.

REPLACING TWO-WAY DISPATCH RADIO

One of the staples of police and emergency medical technician work is communication via two-way radio. Unfortunately, this method is imprecise and error-prone. Directions and information exchanged between unit and base station must often be repeated. While one unit is talking with dispatch, all other units must wait, no matter how urgent their communication. A better solution would be a wireless data system that would handle air radio traffic. The city of West Jordan, Utah, developed a mobile communication system for car-to-car messages and dispatch. A proprietary network was considered and rejected because of cost. Why not use the national CDPD network instead? Thus, a Premiere Mobile Data Terminal Talk-Thru RF server was used to compress and encrypt communications. Connectivity was provided by AT& T Wireless Packet Data Service. Routed through AT&T InterSpan Frame Relay service, officers can communicate information in real time without going through dispatch. The city of West Jordan estimates that deploying three officers equipped with CDPD devices is equivalent to one additional officer on the street.

Emergency medical service has always depended on the two-way radio communication model. In the late 1990s, the city of Calgary, Alberta, deployed a CDPD network provided by Telus Mobility with GPS. By glancing at a screen, dispatchers can discern ambulance position using mapping software. Ambulance personnel can be sent to a site instantly, without waiting for radio clearance and other communication to cease. Addresses can be displayed on a map with accuracy, rather than explained verbally over two-way radio. The result is a deep reduction in ambulance deployment time and ultimately, an improved ability to save lives.

WIRELESSLY EXTENDING A WORKPLACE INTRANET

Rubicon Technologies, a mobile solutions consulting company, relies on an application called Workforce Portal for company operations. The application provides access to time and expense tracking, company events, vacation requests, and other internal forms. Originally developed as a Web-based application with Microsoft SQL Server, mobile capabilities were sought for Workforce Portal's second release. SQL Server Windows CE Edition allowed for server tables to be replicated in individual Pocket PC devices. Moving the application to a mobile front end was not very difficult and gave mobile workers immediate access to needed company data.

SUNDRY SOLUTIONS

At a glance, other successfully deployed mobile applications include the following:

- ✓ **Pocket Hoops:** Pocket Hoops is a digital scouting application that allows basketball scouts to record comprehensive field data. Developed by Infinite Mobility, Pocket Hoops lets a scout view a basketball player's performance and take notes, record voice, employ charts and graphs, refer to statistics, and then upload the information via telephone line right after the game. With many teams looking for their next star player, any technology that gives scouts a timely edge in player evaluation is welcome. Pocket Hoops is deployed on a Pocket PC and is used by the Los Angeles Lakers.
- ✓ **The Maritime Operations Boat Inventory system (MOBI):** The Port of Seattle, the largest seaport in the Pacific Northwest, uses mobile devices to monitor electrical usage by pleasure-crafts and other vessels in the port. The Port Commission chose the Microsoft Windows Powered Pocket PC platform to develop an application called the Maritime

Operations Boat Inventory system, or MOBI. The application uses easy-to-understand check-off boxes and prompts. Users move from vessel to vessel to record electrical usage, and synchronize their data using a cradle connected to a desktop PC.

✓ FieldWorker: FieldWorker is a Pocket PC–based outdoor data collection system specializing in the recording of geological and ecologically significant information. It is used for precision recording of agricultural land use, crop management, storm water inventory, and soil quality analysis. One firm entrusted with mapping and taking inventory on streams and wetlands in Alberni Valley, near Vancouver, used FieldWorker to map all currently unmapped vital areas at a 1:5000 scale. This information can be most valuable in the hands of nonprofit organizations or government officials seeking to protect a fragile ecosystem from overbuilding. Entering stream data coordinates into a Pocket PC while onsite has immeasurably reduced the time required for creating accurate wetland maps.

Summary

We've had a general look at the three most popular mobile wireless technologies: the cellphone, the PDA, and the Pocket PC. In the next chapter, international mobile phone technologies, especially the Symbian OS, will be discussed.

Chapter 4

Wireless Future and the Symbian OS

In This Chapter

In this chapter, we discuss how the future of wireless communication depends on open systems, on devices run on flexible, scalable, open standards. We largely talk about the Symbian OS, formally known as EPOC, developed through an alliance between Ericsson, Nokia, Motorola, and Psion. These companies put their considerable heft behind an operating system developed from the ground up for small telecommunication devices rather than behind one that was a miniaturization of a desktop standard. "Small-is-better" concepts, such as power savings and reusable code, are built right into the basic Symbian OS rather than tacked on as an afterthought. We discuss how the Symbian OS embraces Java, XML, SyncML, and other open standards, as well as how such considerations ensure a healthy wireless future. This same need for flexibility is required on the network level as well. The major wireless communications markets are all beating a slightly different path towards 3G, that vision of the future in which everyone will own and enjoy personal high-bandwidth devices. While the Americas, Japan, and Europe all agree on the need for "always-on" high-speed capacities for data/phones, the three mega-markets are each using slightly differing applied technologies to get there. Each mega-market enjoys a taste of 3G potential by employing some sort of technology overlay on existing 2G systems to achieve some benefits now. But each technology builds on its own unique existing infrastructure, one that will not easily be discarded in order to implement some uniform, worldwide 3G standard. Clearly, the Americas, Europe, and Japan (and other markets as well) will each enjoy differing sets of 3G technologies when 3G devices finally do become available. Thus, it's helpful to invest in wireless infrastructures that are flexible. Whether you are purchasing a phone or developing a phone system, you don't want to be shut out of the future through ill-informed technology investments. That's why in this chapter, we take a good look at the open architecture of the Symbian OS, some of the phones it supports, and what those open standards can mean for you.

- ✓ The IMT-2000 initiative
- ✓ Migration to G3 in North America
- ✓ Japan and the I-phone
- ✓ Migration to G3 in Europe
- ✓ The Symbian OS
- ✓ Symbian devices

An International Look at the Wireless Future

Later chapters define and describe the basic telecommunications technologies and the agencies that oversee their implementation. In this section, we discuss what the telecommunications industry is doing to give a glimpse of the future of wireless.

IMT-2000 Origins

Key to this is the GSM, or Global System for Mobile Communication. GSM is the most widely used digital cellular phone system in the world. It was originally defined as a European network standard that would support voice, data, and cross-border roaming. However, GSM is present in 160 countries and is one of the world's main 2G digital wireless standards. GSM is a TDM system, implemented on 800, 900, 1800, and 1900 MHz frequency bands.

In 1992, an international group of telecommunications industry officials and technology watchers set forth an initiative that would unite the world under a single telecommunications standard. The initiative, eventually dubbed IMT-2000, took its name from the targets it prescribed: that by the year 2000, telecommunications systems offer users 2 Mbps bandwidth for their devices (2000 Kbps) and that governing bodies set aside frequencies in the 2000 MHz range for this purpose. Although none of these recommendations or goals was met in the literal sense, the ITM-2000 initiative continues to push for advances and make recommendations.

3G Requirements

A set of third-generation standards, goals that device manufacturers and service providers should work towards, was defined. These standards, called 3G, establish the following specifications. For a device to be called "3G," it must do the following:

- ✓ Ideally provide data speeds of 384 Kbps. At this speed, broadcasting video is possible.
- ✓ Send data via packet technology, allowing the delivery of data in small numbers of bits to be reassembled on the receiving end. Packet data transfer is a much more efficient way to utilize bandwidth.
- ✓ Provide "always-on" capability, allowing users to be notified immediately of e-mail or incoming messages.
- ✓ Provide auto-location — the network would automatically know your location via Global Positioning satellites.

Rather than simply waiting for manufacturers and service providers to produce these magic gadgets outright and to continue to promise the public of their imminent arrival, governing bodies began creatively allotting bandwidth and overlaying technologies to allow a taste of those 3G benefits now.

Migrating to EDGE

In North America, much digital cellular service is provided over D-AMPS, a digital upgrade of the original cellular system on this continent. Implementing 3G benefits, such as providing users with a landline quality voice or high-speed data connection, usually requires frequencies out of D-AMPS range. One such solution is called EDGE, or Enhanced Data Rates for GSM and TDMA/136 Evolution. EDGE can be deployed in the 600 kHz range, making it attractive to D-AMPS administrators, and does not require implementation over CDMA. EDGE provides high data rates, but as a TDMA technology, it will never be upgradeable to a more world-popular cdma2000 or wideband CDMA technology embraced elsewhere.

Japan and I-Phone

We want to consider Japan for a moment and see how they are moving towards greater bandwidth and 3G standards.

Japan has been enjoying Web browsing, instant messaging, and auto-notification of e-mail for years. In Japan, tens of millions of users access the Internet through i-Mode. Users browse content-rich Web sites offering mobile banking, taxi timetables, horoscopes, and other personalized data on cellphones with slightly larger screens than are common elsewhere. Japan's specialized i-Mode phones are "always on." Web access is quick. The i-Mode user can view thousands of Web sites created using cHTML, a subset of HTML. Data rates in Japan are currently 9.6 Kbps, but Web pages, though content-rich, are very small.

A single company, NTTDoCoMo, launched and popularized the i-Mode phone. The same company maintains the i-mode gateway that transmits all i-Mode Web data in Japan. To advance towards G3, NTTDoCoMo has the luxury of obtaining new spectrum and deploying it with hot, new technology. Japan, then, is building wideband CDMA (W-CDMA) networks and hopes to provide such benefits as two-way 384 Kbps automotive connections for remote-control driving.

However, since Japan's system is based on a CDMA technology, you could never upgrade your non-CDMA networks to W-CDMA, no matter how great the bandwidth advantage, unless you built an entirely new infrastructure.

GSM/GPRS

The most common GSM upgrade is GPRS, or General Packet Radio Services. Used in Europe and other GSM networked areas, GPRS adds packet-switching protocols and shortens the setup time for ISP connections. Maximum GPRS data speed is 171.2 Kbps, but rates of 20 to 30 Kbps are more frequently seen. As a packet-switched technology, GPRS lays the foundation for "always-on" devices and the capacity to charge a user for data transmitted rather than time spent online.

Upgrading GPRS systems to EDGE generally requires only a change in modulation, not a major expense. However, upgrading from EDGE to a true 3G technology like W-CDMA would be costly.

The Symbian OS

The pros and cons of the applied technologies mentioned in the preceding sections are what ISP directors mull over as they wander through circuit closets trying to find a way to provide bandwidth for everybody's newest toys. How do you give the people what they want and not go broke doing it?

Individual users or communications officers in charge of wireless purchases for an enterprise have little control over how 3G will look. But you can buy wireless devices that have a better chance of adapting to what the future holds. For that reason, this section discusses the Symbian OS.

The Symbian OS, deployed on data phones by Ericsson, Psion, and other manufacturers, was designed with small wireless in mind. It differs greatly from the large proprietary approach of the Microsoft OS and the efficient but somewhat underpowered Palm OS. Symbian was designed to be a true real-time OS for cellphones and PDA/phones. It reuses code, prevents memory leaks, and supports common standards, such as Java and C++. The Symbian OS consists of the following:

- ✓ A multitasking, multithreaded core
- ✓ A user interface framework
- ✓ Data service enablers
- ✓ Application engines and integrated PIM (personal information management) functionality
- ✓ Wireless communications protocols

Symbian OS Features

In 1999, Symbian shipped Version 5 of its OS, geared towards devices with 640 × 240 screens. The Ericsson R830 Smartphone, the first Symbian OS phone, was released in September 2000. Symbian Version 6 deployed Symbian's open architecture and powered the first open Symbian OS phone, the Nokia 9210 Communicator.

Key features of Symbian OS 6 include the following:

- ✓ Contact management
- ✓ Messaging
- ✓ Wireless telephony
- ✓ WAP browsing
- ✓ All types of Internet mail, including POP3, IMAP4, and SMTP

CONNECTION NOT NECESSARY

The Symbian OS was developed to maintain core applications even after the network connection is lost. Data-based applications that are in use during connectivity do not shut down or become unusable when the connection is lost. And very importantly, Symbian is designed to support forthcoming standards, such as W-CDMA. The network stacks are designed to work independently of applications run on top of the OS. A new protocol will not disrupt the application, but rather, enhance it. The operating system does this by providing a rich set of APIs (application program interfaces). These allow developers to "call" other systems in the program and activate a feature set.

SYMBIAN AND JAVA

Symbian licenses its OS code to application and hardware developers. As an open standard, Symbian gives Java developers access to core features of the OS, such as calendaring, telephony,

I/O, and power monitoring. Symbian provides Java implementation within the Java 2 Platform, Micro Edition (J2ME) framework. Developers can utilize the Personal Java Application Environment, which is a Java platform optimized for mobile devices. Mobile phones running the Symbian OS are not dummy terminals with microbrowsers, but rather are hand-portable computer systems. Thus, application development is encouraged. Symbian provides training, tutorials, and other support materials to developers interested in creating applications for Symbian-based devices.

OBJECT-BASED OS

Along with C++, Symbian was developed with OPL, similar to BASIC, specializing in rapid application development. Symbian uses OBEX Object Exchange, a set of protocols allowing objects, such as vCalendar scheduling and vCard contact information, to be portable and exchangeable between other Symbian devices. Developers can create applications that utilize these object-oriented features.

NON-PROPRIETARY SYNCHRONIZATION

In a move away from proprietary desktop synchronization solutions, Symbian has incorporated SyncML into its OS. SyncML is an industry standard for data synchronization of all types. The specification is especially useful for synchronizing contact information, agendas, and to-do lists. SyncML was designed for the less-than-optimum file transfer conditions encountered by wireless devices: slow speeds and unreliable connections. SyncML can exchange messages and transmit data over any network and on diverse protocols. SyncML can work with personal object data, such as vCalendar scheduling objects and vCard business card objects. It also transmits relational data, such as tables, as well as XML documents.

Symbian Phones and PDAs

The following sections include current Symbian OS phones and PDAs. Note that only GSM 1900 modules can be used in North America. You'll notice lots of diversity within these devices. They are not at all subtle spins on the same basic design. Note that the PDAs have larger screens than most North American models.

THE NOKIA 9290 COMMUNICATOR

The Nokia 9290 Communicator (Figure 4-1) is a fully integrated mobile terminal that combines phone, fax, e-mail, calendar, and imaging. It also has an active-matrix screen. Other features include:

- ✓ Optimized for mobile Internet WAP and HTML browsing
- ✓ E-mail and SMS applications
- ✓ Integrated PIM systems
- ✓ Full color, nearly 1/2 12-bit color (4,096 colors)
- ✓ Complete mobile multimedia
- ✓ View Word documents and PowerPoint slides in full color
- ✓ Open OS design
- ✓ GSM 1900

Figure 4-1: The Nokia 9290 Communicator
(photo courtesy of Nokia)

ERICSSON R380 WORLD SMARTPHONE

The R380 from Ericsson (Figure 4-2) combines a full-featured mobile phone, personal organizer, and address book into a single unit. The unit provides advanced messaging, calendaring, and mobile Internet, including WAP with enhanced WTLS security class 1 and 2. Features include:

- ✓ Touch-sensitive, backlit screen
- ✓ Browse WAP pages and full Internet
- ✓ Unified e-mail, SMS, and fax support
- ✓ Handwriting recognition
- ✓ Built-in infrared modem
- ✓ Contact management, including agenda, notepad, and event alarms
- ✓ 900/1900 GSM

Figure 4-2: The Ericsson R380 World Smartphone (photo courtesy of Telefonaktiebolaget LM Ericsson)

DIAMOND MAKO

The Diamond Mako (Figure 4-3) is a half-VGA monochrome unit with a 6.8-square-inch screen area. Infrared port and docking station are also available. Features include:

- ✓ 7-ounce multifeatured organizer
- ✓ 16MB RAM
- ✓ 480 × 160 touch-sensitive screen
- ✓ Spreadsheet and word processing software, as well as other office applications
- ✓ PIM tools
- ✓ Synchronization software
- ✓ Optimized for secure WAP browsing
- ✓ 1900 GSM

Figure 4-3: The Diamond Mako (photo courtesy of SONICblue, Inc.)

NOKIA 7650

The Nokia 7650 (Figure 4-4) is the first 2.5G Symbian phone. It features an integrated digital camera, a photo album, and an image viewer. Features include:

- ✓ For Europe/Asia Pacific
- ✓ HSCSD and GPRS for high-speed data transmission
- ✓ Message optimized for quick sending of photos
- ✓ Multimedia messaging and e-mail
- ✓ WAP browser
- ✓ Infrared and Bluetooth connections
- ✓ PIM features
- ✓ Open OS design

Figure 4-4: The Nokia 7650 (photo courtesy of Nokia)

PSION PDA SERIES 5MX

The Psion PDA 5mx is a palmtop nearly seven inches wide and more than three inches deep, weighing a little more than twelve ounces. Utilizing the 32-bit RISC-based ARM710T RISC CPU, the 5mx runs at 36MHz and on two AA batteries. The CompactFlash slot can accommodate Psion's 20MB memory disk for expanded storage. Features include:

- ✓ 16MB RAM expandable by CompactFlash (CF) card
- ✓ 640 × 240 backlit touch-sensitive screen
- ✓ Built-in digital recording microphone
- ✓ Messaging suite
- ✓ PIM tools
- ✓ Synchronization
- ✓ Mobile office suite of applications

REVO PLUS

The Revo Plus includes a messaging suite of applications and is optimized for Internet connectivity and enriched WAP browsing. Features include:

- ✓ 7-ounce full-function organizer
- ✓ 16MB RAM
- ✓ 480 × 160 touch-sensitive screen
- ✓ Mobile office suite of applications
- ✓ PIM tools
- ✓ Synchronization tools

PSION 618C

The Psion 618C is the Chinese version of the Psion Revo, with over 13,000 Chinese characters. Available in Taiwan and Hong Kong, it includes a Chinese-English and English-Chinese dictionary. Features include:

- ✓ Five input methods, including Zhu Yin, Chang Jie, Pin Yin, and handwriting recognition
- ✓ 7.1-ounce full-featured organizer
- ✓ 480 × 160 touch-sensitive screen
- ✓ 8MB RAM
- ✓ Mobile office suite of applications
- ✓ PIM tools

- ✓ Messaging suite including e-mail and Internet access
- ✓ Synchronize with mobile phones for e-mail and Short Message System messages.

Summary

These first four chapters have explored the world of small, portable devices. In Chapter 5, we examine wireless LAN devices, basic specifications, and local network setups of various types.

Chapter 5

Wireless LAN Devices and Specifications

In This Chapter

In this chapter, we discuss various specifications for wireless LANs. We cover the 802.11 specification in some depth and describe the corporate and office LAN connection as well as the Home RF specification. We discuss connecting via Bluetooth and also look at a few of the most significant WLAN devices. Chapters 6, 11, and 12 also discuss WLAN technology in a number of applications.

- ✓ LAN strengths
- ✓ LAN deployment options
- ✓ The home LAN
- ✓ The 802.11 specification
- ✓ The Bluetooth specification
- ✓ Exceptional LAN devices

What LAN Provides

The wireless LAN in the workplace provides four portable portals (see Figure 5-1). Anyone wirelessly connected using a laptop, PDA, or Pocket PC can connect to the following:

- ✓ **The user's own desktop computer:** The wireless user can access all the files on the desktop as if she or he were sitting at the desk.
- ✓ **All shared drives:** These include network drives containing company forms, installable applications, shared applications, and so on.
- ✓ **The company intranet:** This may be a work portal with company policy updates, late-breaking news, access to company standards, vacation request forms, and so on.
- ✓ **The Internet and e-mail:** The wireless user can connect to the Internet as long as the network itself can do so.

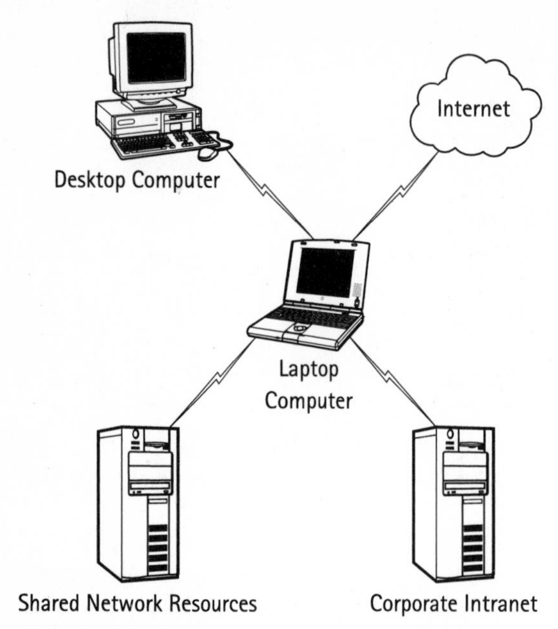

Figure 5-1: A wireless LAN (WLAN) in the workplace

LAN Ranges

Generally speaking, a wireless LAN will connect devices within 100-200 feet of their access point. Walls and electrical interference can adversely affect LAN coverage. However, thoughtful placement of multiple wireless access points (wireless network bridges) can extend a wireless LAN throughout an entire corporate office. Wireless access points are relatively inexpensive. Because they operate similarly to cellphones accessing cells while roaming, wireless devices can move between the ranges of various access points and not lose the connection.

Most wireless LANs use a frequency very close to those used by commercial microwave ovens. Thus, if you attempt to access the network wirelessly while near a microwave oven in operation, you will probably not be successful.

To share connection resources with another office or off-site plant (Figure 5-2), simply set up a wireless point-to-point network with a wireless bridge. By using microwave transmitters and receivers, these “line of sight” connections can bridge distances up to 20 km.

Wireless users can access company resources at any telephoning distance using a virtual private network (VPN). A VPN is essentially a wireless gateway that enables users to safely access e-mail and other networked resources from any distance via an encrypted connection.

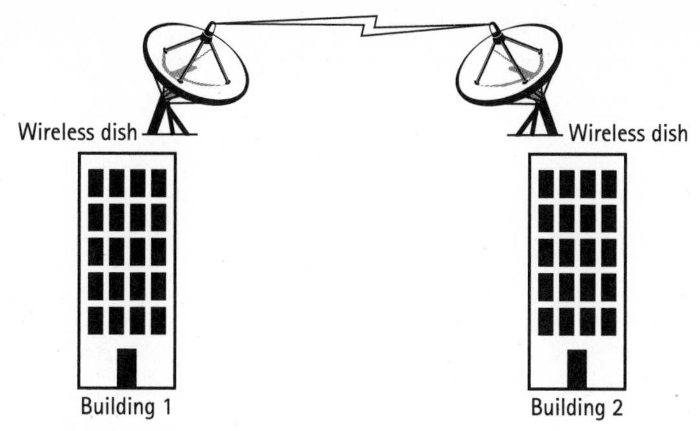

Figure 5-2: A wireless point-to-point network

Mixing LAN and Wired Networks

Generally speaking, companies seldom maintain a completely wireless network. Regardless of convenience or savings, the cost of uprooting existing Ethernet connections and replacing all connectivity with wireless would be far too great. In the last several years, wireless data speeds using the 802.11 specification approach 11 Mbps, certainly fast enough for the most robust networked applications. Still, rather than replace an existing LAN, companies set up a wireless LAN to accommodate mobile employees. These could include staff who like to roam the halls with their notebooks and PDAs and still stay connected or core committees who want to want to have that 1:00 p.m. product development meeting in the cafeteria and still stay connected. Also, road warriors attending trade shows and conferences benefit from wireless LANs as well as do IT heads who want to provide general access to a new piece of equipment without drilling holes in walls and laying down more wire. It's quite easy to extend an existing LAN wirelessly, as we discuss later in the chapter.

The Home LAN

In the home, wireless networks are used to connect multiple computers, allowing the sharing of printers, scanners, and so on. While few home users would relish drilling holes in floors and walls to accommodate a wired network, sharing computer resources without wires sounds downright enticing to many. The cost of setting up a wireless home network has fallen to well below $500, depending on the number of devices and client machines involved.

Wireless LANs are great for the home. Without wires, a computer upstairs can print a special document on the expensive mega-printer elsewhere in the home or offload files from a crowded laptop onto a desktop machine that has space to spare. But primarily, home networks are used to share a single high-speed Internet connection. Properly connected, one high-speed DSL or cable connection can be accessed by all computers connected via wireless LAN. The next couple sections of the chapter discuss how sharing a home Internet connection works.

LAN SETUP FOR MULTIDEVICE INTERNET ACCESS

The first step in setting up an Internet access-sharing wireless home network is to set up a wireless hub or router. This device plays three roles:

- ✓ It provides a physical connection to the Internet source — the cable, DSL line, or phone line.
- ✓ It has an IP address, which is a numerical identifier allowing the computer to be recognized by the Web host and, thus, connect to the Internet.
- ✓ It transmits the wireless signal to other wireless devices within range. This it does via an antenna. The wireless hub/router shares Internet data with other wireless devices and facilitates a network connection between them.

SHARING AN IP ADDRESS

Suppose you have one IP address, but several computers that want to all get on the Internet. If each computer requires an IP address in order to surf the Web, don't all computers need their own access? Not necessarily. The TCP/IP layer has set aside a class of IP addresses for internal use only. These addresses are for devices that are not exposed to the Internet directly, but rather receive a special IP address from this approved set. This is done via a technique called *Network Address Translation* (NAT). Your home computer's IP addresses — those connected via the wireless router — will be something like 192.168.1.0, and the next computer in your LAN will have an IP address something like 192.168.0.2, and so on. For example, if you have a home LAN connecting several devices to the Internet and sharing data with one another, the hub (or router) would have a unique IP address that connects to the Web. The connected devices would be assigned an IP address from a NAT set of addresses. These addresses are set aside specifically for devices that will not access the Internet directly but via a hub or router as indicated.

You're thinking, "But what about duplication? I thought all IP addresses had to be unique?" Yes, but this special set of addresses is set aside for computers that will be exposed to the Internet only via a hub/router, as described here. So, turn on more wirelessly enabled computers within connection range, and each will be instantly assigned an IP address and can surf the Web.

When you install your wireless network card, make sure that you check the "Obtain an IP address automatically" option. In Windows computers, this is found on the TCP/IP Properties dialog box of the Network dialog box when setting up your device (see Figure 5-3).

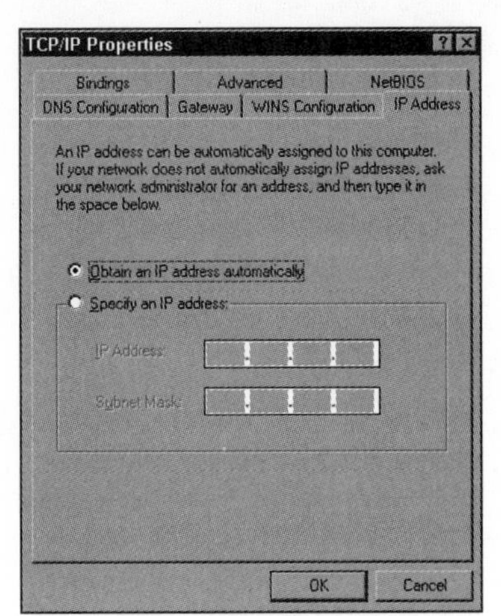

Figure 5-3: From your Network settings, the IP Address tab in the TCP/IP Properties dialog box is where you can choose to obtain a network address automatically.

Then, of course, all computers in your wireless network have access to shared hard drives, printers, and other networked devices. Each computer can have the option of making its hard drives, specific folders, and printers available to other computers in the LAN. In Windows PCs, to share a hard drive, choose My Computer and right-click on that hard drive's icon. Choose Properties and click the Sharing tab. You can choose to allow complete access, password-restricted access, or read-only access.

Peer-to-Peer LAN

The connectivity described so far in this chapter involves several devices clustered around an access point or roaming between several access points, if necessary to stay connected. However, in the absence of access points, wireless-enabled devices that are within proximity of one another can share files, even if there is no access point. They can connect peer-to-peer (this is true using the 802.11 wireless specification). Peer-to-peer is a method for computers to share Internet access, hard drives, files, printers, and other resources without using a hub or router (see Figure 5-4). This solution is fine for a very small number of devices. Connecting more than three or four computers this way usually results in decreased access speed.

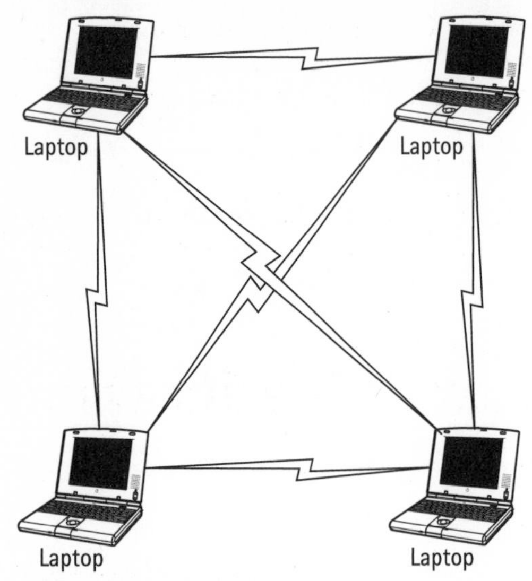

Figure 5-4: Laptops connected through a peer-to-peer wireless network

The Home LAN Advantage

Home wireless systems have great potential. The following devices and applications are possible for homes that have multiple WLAN access points and have purchased equipment utilizing that technology. In a wireless home system, the devices discussed in the next few sections can be placed anywhere around the home where there is adequate signal to maintain a good connection.

Home Security and Monitoring

Security systems are available that trigger audible, remote alarms in the event of an intruder. Wireless security cameras can be placed in front of entrances, in children's rooms, or at any location you'd like to monitor. Camera output can be displayed on a central computer in the home. While on the go, connect the feed from those cameras to a Pocket PC or laptop with a wireless WLAN card.

Environmental Control

Lights and heating can be controlled by a central remote. Monitor environmental settings not only from your home PC or laptop but also on the go by using a WLAN card that provides you with a VPN or other network connection to your home system. Control environmental settings remotely so that the house is lighted and warm when you arrive rather than governed by a clock.

Conditional Device Management

Coffeemakers can be triggered by an Internet-based alarm clock. The alarm clock can self-adjust based on local traffic conditions reported via Internet news channels. In the works are

Internet-connected refrigerators that keep track of food use. When you put food in the fridge, pass a bar code over each item. The refrigerator notes food items and quantities. When items run low, a shopping list is printed or sent via Internet to a local grocer who delivers replacements.

Entertainment

MP3 players and other multimedia entertainment equipment can send music to speakers placed anywhere in the house within wireless range. Music and other media playback and reception devices can reside in one central location, while speakers and monitors are mobile.

802.11 and Related Network Specifications

Before we discuss WLAN devices, a short history of the 802.11 and related specifications is useful.

802.11 Standard History

The United States (the FCC in particular) is rather generous with spectral wavelength. Rather than holding back huge portions of the spectrum for a yet-to-be agreed upon standard, the FCC is quick to set aside bandwidth for public use. Thus, in 1985, the FCC designated the 2.4 GHz microwave band for unlicensed use. The 2.4 GHz range is rather crowded, being already used for industrial, scientific, and medical applications (ISM). Also, microwave ovens and other electrical devices operate in the 2.4 GHz range. Nonetheless, in 1997, an engineering body known as the IEEE 802 committee developed a framework for wireless networked communications using the 2.4 GHz band.

IEEE stands for Institute of Electrical and Electronic Engineers, a North American professional body (see `http://standards.ieee.org/`).

Most WLAN systems use spread spectrum technology, such as 802.11. By spreading the data across various sub-bands, the data is more secure because it is less easy to hijack. Also, if data is spread across sub-bands, at least some portion of the data is likely to reach its destination, and the recipient system can simply request a resend of the portion that didn't arrive. Spread spectrum, then, sacrifices efficiency and speed in order to gain security and reliability.

So, WLAN data is encoded across spectral sub-bands before it is sent. Initially, the 802.11 specification allowed two methods of encoding:

- ✓ **FHSS, or Frequency Hopping Spread Spectrum:** This method spreads that data across 75 single-megahertz subchannels, continuously hopping between them. The idea was that if one channel were blocked for some reason, data would get through one of the others. The initial deployment of the 802.11 spec used this relatively inexpensive method of encoding.

- **DSSS, or Direct Sequence Spread Spectrum:** This method divides the band into 14 overlapping 22 MHz channels, using each sequentially. Since larger amounts of data are sent simultaneously, DSSS increases the potential data rate from 1 to 2 Mbps. However, before DSSS encoding could be used, the problem of increased power consumption had to be solved.

The Birth of 802.11b

Once deploying 802.11 with DSSS encoding became feasible, the specification was revamped and released again, this time with more efficient modulation. Known as 802.11b, wireless networks were then capable of transmitting 11 Mbps, just a tad faster than the original Ethernet spec, which delivered data at 10 Mbps. Such a boost in speed was a real face-lift for wireless 802.11. With this improvement, wireless LANs began to be considered a viable alternative to cabled Ethernet connections for small local networks.

The 802.11 Layers

A network specification implements a set of controls over at least one network layer. The 802.11 spec places specifications on both the physical and the medium access control (MAC) layers. A physical layer in a network defines the signaling and modulation characteristics of data transmission. The 802.11 physical layer makes provisions for data rates as well as encoding type (FHSS or DSSS). The MAC layer is where provisions are set to avoid data collision. When you have multiple data streams transmitting at once (and the more the better, for maximizing bandwidth efficiency), collision avoidance is a necessity. The 802.11 standard specifies a Carrier Sense Multiple Access with Collision Avoidance (CSMA/CA) protocol. The MAC layer governs how a node responds to packet traffic.

What is ideal is to have a packet-transmitting node wait for an "all clear" before sending a packet of data. However, if all nodes sense a channel is in use and wait the same interval before sending again, each node will again attempt to send and again find the channel busy — ad infinitum. The idea, then, is to vary the waiting intervals so that each node has a unique window of opportunity to send its packet.

Where They Get the Name "802.11"

The IEEE Standards Association uses a numerical system to identify its initiatives and standards. 802 represents local area and metropolitan area network standards. There are several LAN/MAN standards. 802.5 represents LAN/WAN standards related to the Token Ring topology. 802.3 designate Ethernet standards. And finally, 802.11 represent wireless LAN/WAN topology standards. 802.11b represents the 802.11 revision in which DSSS encoding began to be used.

So, here's how the engineers solved this problem in the 802.11 spec. When a node receives a packet to be transmitted, the node determines if data is currently transmitting on the channel. If the channel is not in use, the node sends the packet. If the channel is busy, the node chooses a random "back off factor," which determines how long the node should wait before attempting transmission again.

However, to avoid needless waiting, the node reduces its waiting period during instances in which no other traffic is detected. So, since the chances of two nodes choosing the same "back off factor" are small, the node is likely to be able to transmit its packet very shortly. Thus, the 802.11 spec is able to provide fluid data flow with minimal collisions.

Faster Wireless: 802.11a

Although not many devices exist for it yet, 802.11a is a specification that holds much promise. This specification has the potential to run at speeds of 54 Mbps, which would allow you to use it to download large files, run applications, and display multimedia content. The 802.11a standard operates in the 5 GHz frequency range, and the FCC provided a much larger piece of the spectrum for unlicensed operation of the block (which would include 802.11a) than it did for technology using the 2.4 GHz range. This broader allotment of bandwidth allows the specification to utilize an encoder that supports faster file transfer and better performance than the spread spectrum technologies (FHSS or DSSS). The encoder, known as COFDM, breaks a used segment of bandwidth into 20 MHz subchannels that can pull much more data through at higher speeds than 802.11b is capable of. 802.11a performance is also improved by the relatively underpopulated 5 GHz frequency region, resulting in much less interference.

The ability to run eight high-bandwidth channels simultaneously by using 802.11a makes transmitting DVD signal and other bandwidth-hungry tasks a viable possibility. Currently, Proxim offers both PC cards and an access point for the 802.11a spec. On these devices, streaming video data runs well, and large files in the 100MB range transfer quickly without incident.

Since the 802.11a specification runs in a higher frequency range than 802.11b, coverage area may be shorter, although 802.11a's more-powerful technology mitigates some of that adverse effect. Still, it's just a fact that data travels shorter distances at higher frequencies.

The Bluetooth Specification

Bluetooth is a wireless specification for a discrete radio connection between all kinds of devices. Operating in the 2.4 GHz license-free band, devices using the Bluetooth specification can connect with one another by simply being in range. Place a Bluetooth printer in a room, turn in on, and start printing. Phones with Bluetooth headsets are truly hands-free.

Bluetooth technology was first developed in 1994 when Ericsson Communications began working on wireless connectivity between cellphones and various accessories. A small-form, low-power, and low-cost radio solution developed, one that could be utilized in all kinds of electronic devices. A Bluetooth-enabled device would "know" which devices to connect with, when possible. (See Figure 5-5.) MP3 players would wirelessly connect with headsets and stereo speakers. The immediate application of Bluetooth technology is aimed at connecting smart phones, PDAs, and computers. Bluetooth is inexpensive, fast, and small. It provides connectivity of up to ten meters with a throughput of 720 Kbps. Bluetooth is housed on a single-chip radio that costs about $5, and most manufacturers find this a very affordable technology and are eager to apply a Bluetooth solution of their own. Bluetooth currently supports device link ranges of up to 10 meters. Improvements in transmission power will soon increase that range to nearly 100 meters.

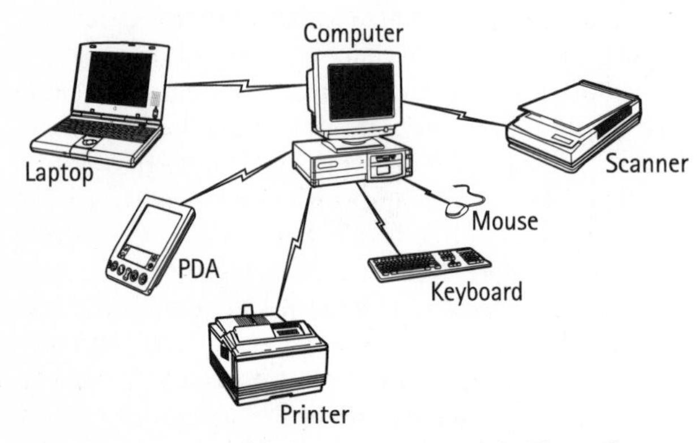

Figure 5-5: A range of device types connected via Bluetooth

Bluetooth devices form natural *picocells,* or peer-to-peer connection rings, in which all applicable devices become aware of each other and begin sharing data simply by coming into range. However, Bluetooth aggressively applies a frequency-hopping technology that creates interference with other nearby devices that also use the 2.4 GHz band. And while Bluetooth devices can change frequencies up to 3,200 times per second and work just fine in such an environment, other devices vying for the same bandwidth will have their performance affected by the nearness of Bluetooth devices. Additional hubs and routers may be required to overcome the performance hit, or you can simply move the Bluetooth devices a few feet out of range, if possible.

Home RF

Another network spectrum that will probably increase in popularity as more devices are made to use it is Home RF. Home RF can run home devices locally, just like 802.11b, but can support up to eight high-quality voice connections, eight streaming video or audio sessions, and shared network resources and Internet access. Home RF is the only specification discussed in this chapter that is capable of voice quality in league with a land-line phone, otherwise known as toll quality. Home RF also supports the full set of phone-based features, such as call waiting, call forwarding, distinctive rings, and so on. Home RF is based on the original FHSS 802.11 spec. Because this spec involves spreading the signal across multiple frequencies to insure data transfer, transmission will generally be slower than the 802.11b spec would allow.

Network Choice Considerations

When picking a wireless network, you have to choose only one of the specifications discussed here and build your home or enterprise wireless network around it. You cannot mix them. With the exception of 802.11a, every network spec discussed vies for the same crowded 2.4 GHz range. Even if, for some reason, you wanted to invest in access points or hub/routers for two specifications, you certainly would not want to suffer the connection losses that would result as they dual with each other for bandwidth.

The general wisdom of the moment is this: Home RF supports many features, even though Home RF bandwidth support is not the greatest. However, not that many devices out there currently support it. 802.11a is fast and enjoys a bit of uncrowded spectrum all to itself. However, as of yet, device support is weak. 802.11b is very popular, and many, many home and office devices will run with it. Nonetheless, it is in a crowded spectral area that is getting more crowded with the arrival of Bluetooth devices. Expect more congestion and interference with other electronic devices, such as cordless phones and microwave ovens, that also use the same spectrum spread.

All the advantages and disadvantages of the other specifications ultimately leads you back to considering Bluetooth, which is highly automated, simple, and powerful. Many manufacturers are well into the planning stages of providing Bluetooth devices. Note also that because Bluetooth uses frequency hopping, you can deploy many Bluetooth devices in close proximity without concern for transmission problems. In the meantime, those interested in this technology must wait. But when Bluetooth devices do arrive, the standard will probably become one of those technologies that we use and forget about. The connections will simply work.

Exceptional LAN Devices

In previous chapters, we have mentioned wireless LAN cards. In the rest of this chapter, we get more specific and discuss a handful of some of the more exceptional LAN products, especially those that have a future that looks fairly bright.

The Proxim Harmony 8570 802.11a Access Point

Note the "a" in that product description; the Proxim Harmony uses the 802.11a spec. With this product (see Figure 5-6), together with Proxim's wireless adapters, get ready for speeds about 4.5 times faster than 802.11b transmissions. Multimedia files play back with barely a hitch, including full-screen DVDs and other large file transfer tasks you would not undertake on an 802.11b system without serious time to kill. Although the Harmony 8570 works only with Proxim's AP controller, making it a professional device, it will transmit data from 802.11b devices as well, allowing a comfortable migration to the newer, faster spec. This is important because the year 2002 and 2003 will undoubtedly mark the arrival of many 802.11a devices.

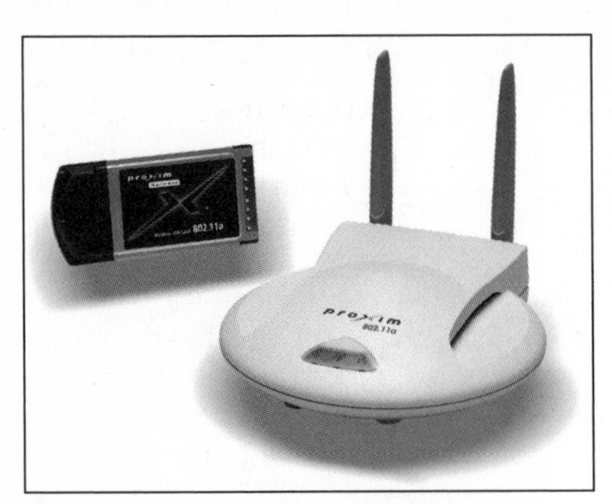

Figure 5-6: The Proxim Harmony 8570 802.11a access point and PC card (photo courtesy of Proxim, Inc.)

The Linksys Instant Wireless Access Point

One of the most popular access point products for both home and corporate use, the Linksys Instant Wireless (see Figure 5-7) is fully 802.11b compliant and Wi-Fi certified, offering up to 128-bit encryption. Extremely easy to install and use, Instant Wireless Access Point extends the reach of wireless devices or provides the initial bridge from a wired connection to roaming wireless devices.

Figure 5-7: The Linksys Instant Wireless access point (photo courtesy of Linksys Group Inc.)

The Ositech King of Hearts Cellular PC Card

This extends your laptop or other PC card-capable device beyond the LAN environment. You can connect to your VPN or LAN via dial-up by using your laptop or Pocket PC and cellphone. The King of Hearts (see Figure 5-8) requires no setup or special plan. It works with your existing connection and uses your cellphone minutes to run a data connection. This is great in a pinch when you need to access the LAN from a distance.

Figure 5-8: The Ositech King of Hearts Cellular PC card (photo courtesy of Ositech Communications, Inc.)

D-Link DI-714 Wireless Broadband Router with 4-Port Switch

This product connects to the Internet via both Ethernet and wireless 802.11b. It has four Ethernet ports, an 802.11b Wi-Fi antenna, an Internet router, and a firewall. Computers connected via the D-Link DI-714 (see Figure 5-9) require no special software. Installation is fast. The device automatically assigns IP addresses to all detected terminals. Of course, make sure that your devices are set up to accept a dynamic IP, and you're off. The product can manage up to 252 users, each with unique internal IPs, but exposes only a single IP to the Internet. Uniquely, though, you can still host Internet services in-house to your connected devices. The DI-714 supports tunneling protocols, such as Point-to-Point Tunneling and Layer 2 Tunneling Protocol. The unit's Virtual Server feature enables you to expose HTTP, FTP, games, and other server types to the Internet, while still shielding the home devices.

Figure 5-9: D-Link DI-714 Wireless Broadband Router (photo courtesy of D-Link Systems, Inc.)

The WIDCOMM BlueGate 2100 Bluetooth Access Point

The WIDCOMM BlueGate 2100 (see Figure 5-10) is the first available Bluetooth wireless access point. BlueGate acts as a gateway for shared Internet access and can link seven Bluetooth-enabled devices simultaneously.

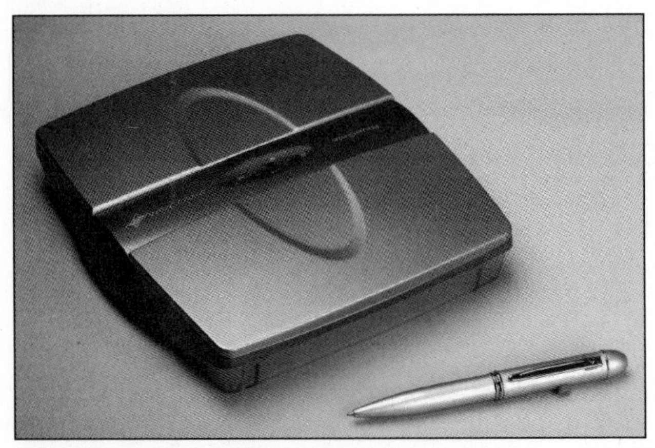

Figure 5-10: The WIDCOMM BlueGate 2100 Bluetooth access point (photo courtesy of WIDCOMM, Inc.)

The Palm SD Bluetooth Card

The Palm Bluetooth Card comes in the tiny SD form factor and provides Bluetooth connectivity for the Palm m500 and m505 handhelds. These cards share data with other Bluetooth devices that are PDA-compatible.

The Xircom RealPort 2 Bluetooth PC Card

To connect with other nearby Bluetooth-enabled devices, just plug the RealPort 2 card (see Figure 5-11) into your PC card-enabled laptop or other PC card-enabled device. As soon as the laptop is within range of another Bluetooth device, data sharing and access can begin.

Figure 5-11: The Xircom RealPort 2 Bluetooth PC card (photo courtesy of Xircom, Inc.)

D-Link DWL-650 Wi-Fi Card

D-Link's DWL-650 802.11b PC card will get your laptop or PC card-enabled device networked quickly and inexpensively. Currently priced at less than $100, the DWL-650 is compliant with all major networking protocols and works well with all current Windows versions. The DWL-650 has an outdoor range of up to 300 meters, automatically switches down from 11 Kbps to lower data rates if conditions require, and seamlessly moves from an ad hoc peer-to-peer mode to an access point connection.

Sonic Blue ProGear Wireless Big Picture Computer

Sonic Blue's ProGear (see Figure 5-12) is a 802.11b portable computer with a touch screen 10.4-inch LCD flat panel display. It looks like a high-tech version of the old child's toy Etch-a-Sketch, but a toy, it's not. ProGear has 128MB RAM, a 5.6GB hard drive, and runs with a Transmeta 400MHz-compatible processor. Using a built-in 802.11b wireless card with embedded antenna, ProGear seamlessly connects to local corporate data (available with a PC Card/CardBus option as well). Geared towards enterprise use and targeting a number of industries (health care, hotel services, education, and automation systems), ProGear is available in both Linux and Windows 98 configurations. Currently, no other Windows versions are supported. ProGear has handwriting recognition software for writing right on the screen and lets you record your voice. The unit has a six-hour battery life and can be used in portrait or landscape modes.

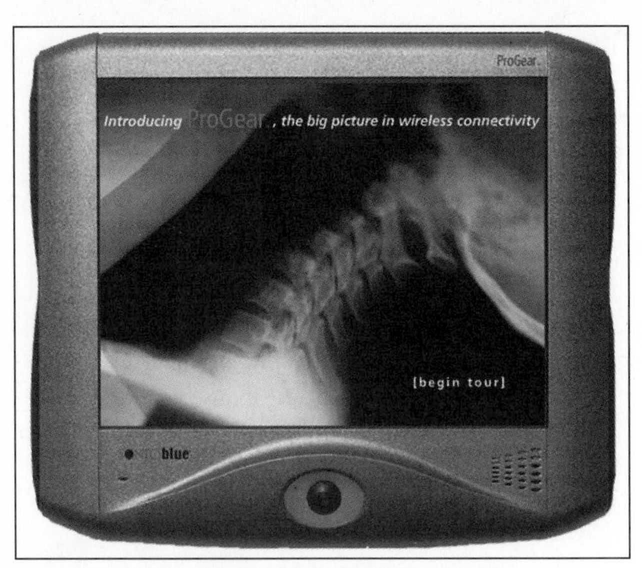

Figure 5-12: The ProGear Wireless Big Picture Computer
(photo courtesy of SONICblue Incorporated)

Summary

You've learned a bit about wireless local area network technology, specifications, and applications. We've talked about various methods for setting up LANs and looked at a few of the available devices. In the next chapter, we discuss a wide variety of wireless devices, great and small.

Chapter 6

Diverse Wireless Devices

In This Chapter

Despite economic downturns, wireless is still an industry that sparks innovative product development. Wireless technology is fascinating to consumers and manufacturers alike; in fact, some wireless innovations are manufacturer-driven — a wireless concept becomes possible from a technical standpoint, and technology is thus wedded to a product. Wireless is still an industry in which a few good ideas and some engineering chops can raise a good bit of investment capital. Investors then hope that consumers catch on, marveling at the convenience, usefulness, or at least the cleverness of some of the gadgets that are tossed their way. This chapter covers a wide selection of wireless devices, focusing particularly on products that supplement those read about earlier.

This chapter discusses wireless communication devices that use a proprietary operating system — PDAs that don't run on Palm, for example, or Pocket PCs that don't run on Windows CE, Pocket PC 2002, or any of its obvious competitors. Selling a frequently used device that runs on a proprietary operating system can be very chancy. If the device doesn't catch on, what about those customers who expect the product to be supported? What about those businesses who went and built an entire corporate communications system on a product that didn't get off the ground in a big way? They'll need technical support, repair parts, and upgrades — entities that manufacturers are not too keen on providing for dying product lines. Still, manufacturers will take chances, providing a feature that "the big guys" overlooked. Wireless communication devices that use their own form factors, run on their own operating systems, and are essentially built from scratch, indeed often include exciting and interesting features. You'll get more RAM or ROM, a digital camera, wireless access thrown in for free, several expansion slot types, a bigger screen, or something that's just plain "cool." We'll look at some of those products in this chapter.

This chapter covers Global Positioning System (GPS) devices, especially GPS driving aids. It also takes a look at the Home RF market, which is a bit of a gadgeteer's paradise. Although consumers may have rejected the self-measuring mixing bowl or the refrigerator that verbally accosts you when you are low on milk, there is still a great deal of interest in Home RF devices and technology. Although most wireless home gadgets have not caught on as expected, consumers like the concept of a secure, wirelessly efficient, convenience-driven home. Home RF is also a very capable specification, competing with 802.11. Several wireless access device manufacturers produce access points, PC cards, and USB adapters for Home RF, rather than 802.11. We'll look at some of those devices as well.

✓ Devices for travelers

✓ Devices that do their own thing

✓ Device accessories

✓ Device display diversity

✓ Device security and connectivity

✓ Devices for the wireless home

GPS Travel Aids

This section describes some of the more advanced GPS products that interface with maps and provide real-time directions while you travel.

TravRoute's Pocket CoPilot 2.0

The TravRoute's Pocket CoPilot 2.0 is a GPS device that connects to your laptop or Pocket PC. Based on satellite tracking, the screen displays a map of your current location as well as your own position in motion. You'll see your position update in real time. Input a destination, and, as you drive, verbal directions will be given towards that destination. If you miss a turn, Pocket CoPilot 2.0 will direct you back on track. Pocket CoPilot 2.0 also responds to voice commands such as "Where am I?" Also provided are a searchable point-of-interest database for travelers, detour/avoid tips, and estimated time-of-arrival. Pocket CoPilot is available as a standalone unit or as a CF card for the Compaq iPAQ, Casio Cassiopeia, or HP Jornada.

Garmin GPS V

Garmin International, creators of the popular e-Trex Summit hiking companion, has now developed a standalone GPS driving aid, the Garmin GPS V. Dashboard mountable or handheld, the unit provides a large, clear interface with easily-seen digits directing you where to turn next. The Garmin GPS V provides audible alert beeps for upcoming turns.

Magellan GPS Companion

The venerable Magellan makes three GPS product lines: handheld GPS products for sports enthusiasts, vehicle navigation systems, and GPS plug-in CF cards for PDAs. One popular product is the GPS Companion for the Handspring Visor (see Figure 6-1). While driving, your location is plotted on downloaded street maps. Your vehicle's speed, direction, and distance to the next waypoint can be displayed. You can also zoom in and out for greater detail or to get more of the big picture.

Figure 6-1: The Magellan GPS Companion (photo courtesy of Thales Navigation)

Unique PDAs and Phones

This section describes some interesting or proprietary takes on the familiar PDA or cellphone devices.

CMC Magnetics CyberBoy

CMC Magnetics CyberBoy (see Figure 6-2) is a PDA running on an open-source proprietary operating system, allowing the development of a unique set of features not often seen in other products. It has a 320 × 240 grayscale display and touchscreen, and it runs with the typical PDA features, such as scheduling, to-do lists, and synchronization options for e-mail.

What sets CyberBoy apart from other PDAs is its built-in camera. The unit captures 640 × 480 images in 24-bit color. When linked via USB to a computer, CyberBoy functions as a Webcam. CyberBoy also plays MP3s with remarkably good stereo quality. While the unit's internal 8MB memory is not large enough to hold much in the way of music, it does have a SmartMedia memory expansion slot. CyberBoy also includes an FM radio with ten programmable presets and a uniquely compressible voice storage technology for recording and storing voice messages. CyberBoy runs on AA batteries or with an optional Lithium-ion battery pack.

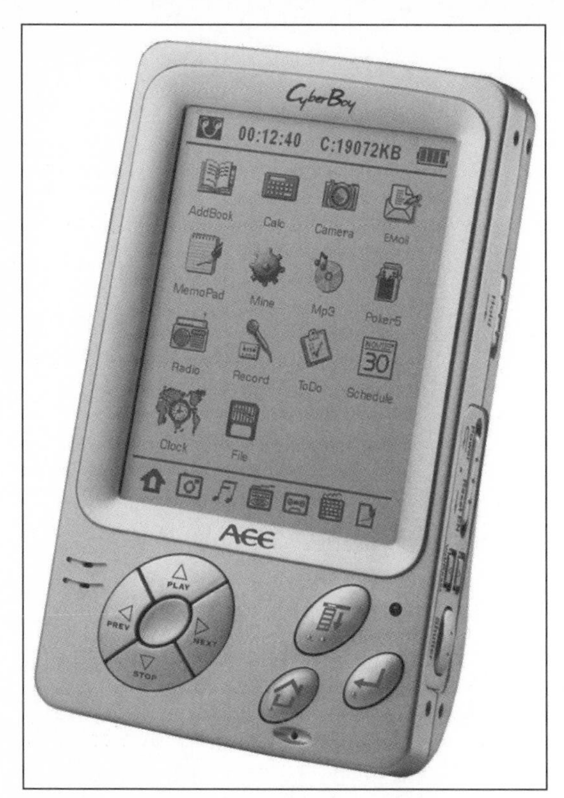

Figure 6-2: The CMC Magnetics CyberBoy (photo courtesy of CMC Magnetics Corporation)

Matsucom onHand

The Matsucom onHand (see Figure 6-3) is a 2.5-ounce water-resistant watch PDA that can store up to 2MB of information. It has a 102 × 64 dot matrix STN LCD backlit display and connects to the PC via serial port or infrared port. OnHand can synchronize with Microsoft Outlook, Schedule +, Palm Desktop, Goldmine, Act!, and Lotus Organizer. Running with a 16-bit CPU, onHand includes standard PDA utilities, such as calendar, address book, to-do list, memo pad, text viewer, and world time. You may enter data on the PC and then synchronize with onHand via Serial or Infrared port and view the data later at your own convenience.

Nokia 7650 Imaging Phone

The Nokia 7650 Imaging Phone is an example of Java and Symbian technology at work. The phone utilizes Multimedia Messaging Service (MMS), enabling the transmission of images, audio, voice, and text in a single message. Initially available in Europe, Asia, and Africa, the phone uses GPRS over GSM 900/1800 frequencies. The Nokia 7650 is Bluetooth- and WAP-enabled. The phone features three data access speeds, depending on the type of task being performed.

Figure 6-3: The Matsucom onHand Wrist Computer (photo courtesy of Matsucom, Inc.)

Keyboards and Input Devices

This section describes keyboards and input devices of various types that add quick data entry to your PDA.

Targus Thumbpad

If you have a problem entering data into your PDA because you can't find a flat surface for typing, the Targus Thumbpad (see Figure 6-4) may be for you. A complete QWERTY keyboard that you support with forefingers and thumbs, Thumbpad attaches beneath your PDA via Serial port. It measures slightly wider than the PDA itself, and weighs in at 1.4 ounces. Typing out short e-mails is no problem, and with some practice, you may even become a pretty fast "thumb-typer." You can use it while standing in line, and no surface is necessary.

Seiko also markets a QWERTY thumb-style keyboard that mounts at the bottom of the unit. Available for most Palm OS PDAs, the Seiko Thumbboard packs more keys than the Targus Thumbpad. The Thumbboard includes very helpful application keys along the top of the device, large four-direction buttons on the lower right, and even an on/off button.

Figure 6-4: The Targus Thumbpad (photo courtesy of Targus, Inc.)

The HalfKeyboard

For mobile users who can spare one hand for data entry but need the other for dialing or carrying, the HalfKeyboard corporation has produced a handy half-keyboard. Measuring less than six inches wide and weighing just under five ounces, the HalfKeyboard plugs into your Palm OS PDA, allowing quick data input.

Rather than squeezing an entire keyboard into an undersized space using microkeys, the HalfKeyboard displays the left side of a standard QWERTY keyboard. To display the right-sided keys, you just press and hold the space bar while typing. As odd as this sounds, users claim to get the hang of it very quickly.

The Fellowes' PDA Keyboard and Case

The Fellowes corporation has a product that satisfies two requirements for serious PDA owners: a durable case and portable keyboard. Most PDAs cannot be bought off the shelf with sturdy cases. (And you should think twice before tossing a $300 computer into your shirt pocket.) Also, most PDA owners, at one time or another, begin wishing for a way to input text quickly. Enter Fellowes' PDA Keyboard and Case for Palm V and Handspring Visors (see Figure 6-5). The unit is at once a snug and sturdy case (it won't increase the PDA's bulk significantly) and a QWERTY keyboard. The PDA fits into a compartment on the left side of the case, while a keyboard appears on the right. The keyboard does not fold out or require extra work room. It remains in the case while you type. The keyboard also includes a number pad and hotkeys for cutting, pasting, and other functions. The hotkeys alone can be real timesavers — a welcome alternative to searching for such functions with your stylus on the Palm.

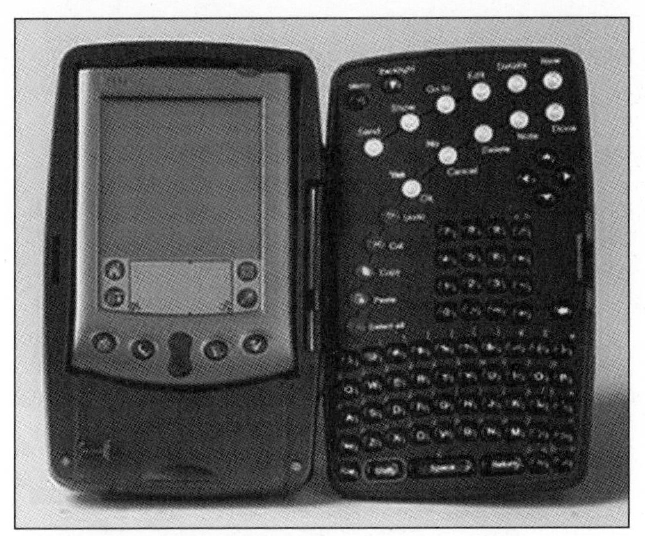

Figure 6-5: The Fellowes' PDA Keyboard and Case (photo courtesy of Fellowes Manufacturing Company)

Seiko SmartPad2

The SmartPad2 (see Figure 6-6) is a fold-over, zippable carrier case for the Palm, Handspring, Sony CLIE, or Handera PDA.. The PDA is inserted on the left side of the case, and a paper pad and tappable QWERTY keyboard are provided on the right. Text or drawings you create on the paper pad are reproduced as text or bitmaps on the Palm. Using the SmartPad pen, you can create charts, freehand text, and drawings of all types. Anything you draw on the paper is saved on the PDA. Later, you can HotSync from PDA to PC and edit your work. Beneath the paper lies an electrical pad that transmits pen strokes to an infrared transceiver. The transceiver data is transferred to the PDA. Also included beneath the drawing pad is a tappable keyboard that inputs text into any of the four primary PDA applications: Address, Memo, To-Do, or Calendar. On the first SmartPad, the Infrared port on the PDA didn't quite line up with the SmartPad's infrared transceiver, but the SmartPad2 took care of that. The SmartPad2 is a hugely popular PDA accessory.

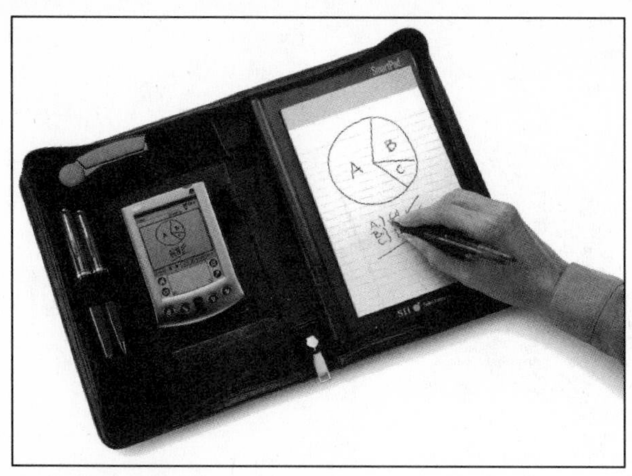

Figure 6-6: The Seiko SmartPad2 (photo courtesy of Seiko Instruments Inc.)

Display Enhancement

This section discusses devices that allow you to send your Palm or Pocket PC display data to a computer monitor or digital projector.

Margi Presenter-to-Go

For presenting full-color slide shows and displaying application screens from a handheld, check out Margi's Presenter-to-Go (see Figure 6-7), which connects your handheld to a VGA screen or digital projector. Connecting via a Type II card slot, Presenter-to-Go can display slide shows from Springboard PDAs (see Chapter 2) or Pocket PCs. The unit includes software for creating slide shows on your PC and then transferring them to the PDA for display from your handheld device. Your presentation becomes truly portable. An infrared 14-button remote control device is included as well. The Margi Mirror software kit will mirror all Palm display activity onto a monitor or digital projector. Screen changes and option selections on the Palm display are shown on the viewing device.

Colorgraphic Voyager VGA Adapter

Colorgraphic's Voyager VGA CompactFlash super VGA adapter is a CF card VGA adapter for handheld PCs. Voyager VGA (see Figure 6-8) enables display output to practically any VGA monitor or digital projector. The unit offers 16-bit color at 640 × 480 or 800 × 600dpi, or 8-bit color at 1024 × 768. Uniquely, both composite and S-video formats are supported. Voyager VGA supports a large variety of Pocket PCs. Your purchase also includes a PC card adapter.

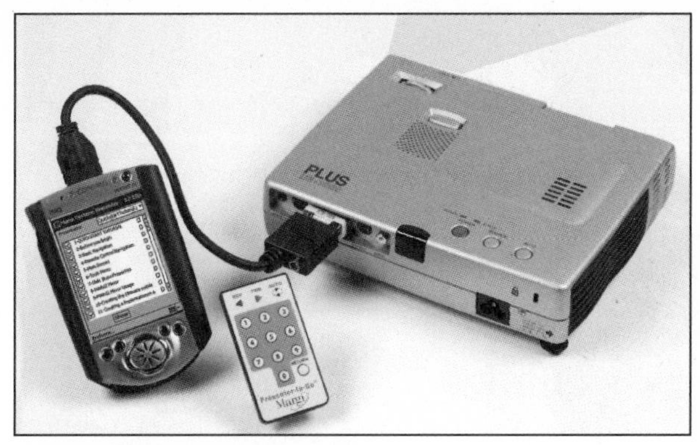

Figure 6-7: Margi's Presenter-to-Go (photo courtesy of Margi Systems, Inc.)

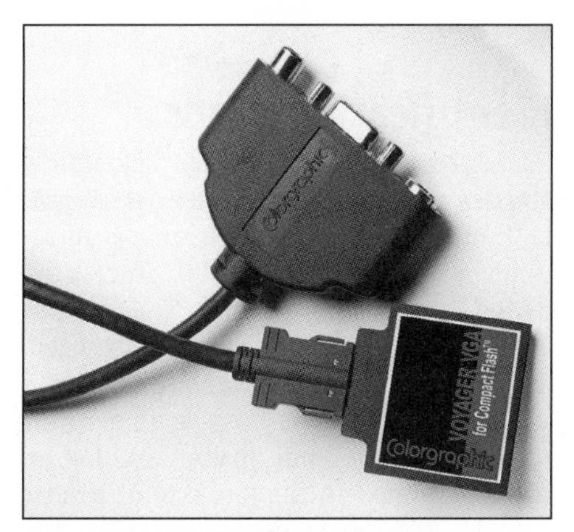

Figure 6-8: Colorgraphic's Voyager VGA adapter (photo courtesy of Colorgraphic Communications Corporation)

Sundry Utilities

This section discusses utilities that provide security or expand the capabilities of your wireless portable device.

Pentax PocketJet 200 Mobile Printer

The Pentax PocketJet 200 Mobile Printer (see Figure 6-9) is a natural printing companion for any mobile computing device. All Windows CE, EPOC 32, Palm, Pocket PC, and BlackBerry OS devices can print with it. Weighing 1.1 pounds (including the battery) and measuring ten inches in length, this is a printer you can truly bring along. The unit never needs replacement ribbons, toner, or ink. Furthermore, the PocketJet 200 prints up to three pages per minute at 200 dpi through IrDA. Pentax provides infrared drivers and printing software with the unit.

Figure 6-9: The Pentax PocketJet 200 Mobile Printer (photo courtesy of Pentax Technologies)

Compaq and Toshiba Fingerprint Identification Cards

The Compaq Fingerprint Identification Technology PC Card provides failsafe pre-logon identification for any Type II PC card-enabled device. The user presses a thumb to the small external pad, and then the device photographs the print and converts thumbprint impressions to an identification digit that allows logon. The initial user registration involves allowing the card to process the thumbprint's mathematical "map" and store the identifying data for future logons. The Compaq device supports Windows versions 95 to 2000, including NT 4.0. Novell NetWare is also supported.

Toshiba also makes a PC-card fingerprint reader (see Figure 6-10), which combines Identix's DFR-300 optical fingerprint reader and BioLogon security software. The Toshiba fingerprint reader supports multiuser environments. Many users can store profile information and log on with the touch of a fingerprint. When used with Toshiba laptop models, the Fingerprint Reader allows remote user logon and the ability to set additional password requirements besides the fingerprint.

Figure 6-10: The Toshiba PC Card Fingerprint Reader (photo courtesy of Toshiba USA)

Connectivity Enhancement

This section discusses devices or software that provide various types of wireless connectivity to your PDA or Pocket PC.

GISMO All-Connection for the iPAQ

Abrandnewworld's GISMO (see Figure 6-11) is a cradle that transforms your iPAQ into a GPRS, GSM, and 802.11b WLAN communicator. The iPAQ sits in the cradle and is completely mobile. GISMO provides telephony, full Internet connectivity, e-mail, and access to company intranets. It is a true Swiss Army Knife connectivity solution for the iPAQ. Note that GSM services are included, so you can have global wireless access. For North American users, GSM 900/1900 is supported. At 5.7 inches high and a weight of 5.25 ounces, the GISMO is rechargeable via AC adapter or iPAQ cradle. The GISMO includes a built-in antenna and headset.

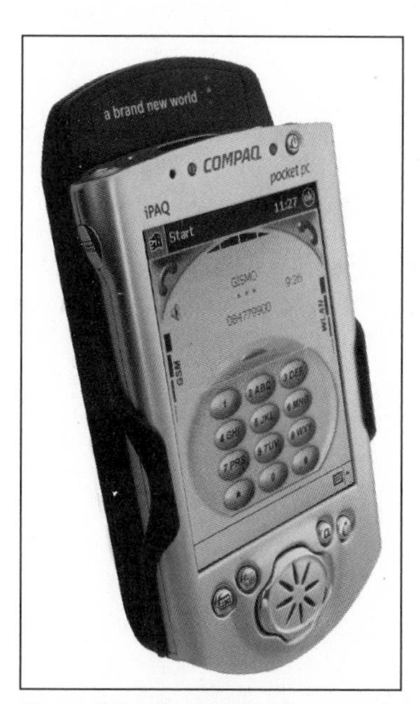

Figure 6-11: The GISMO All-connection cradle for the iPAQ (photo courtesy of abrandnewworld)

LapLink Gold

LapLink is a leader in file transfer and computer synchronization software, specializing in a number of PC-to-PC file transfer, remote access, and control solutions. LapLink Gold is the company's specialized remote access software for IT and mobile professionals. LapLink Gold is a peer-to-peer application, allowing connection to another PC from anywhere. You can run applications remotely, collaborate online via VPN, view text messages and voice chat, as well as synchronize schedules. LapLink Gold is an easy-to-install solution that does not require heavy-duty IT intervention to get up and running. It is a good solution for a mobile professional who needs single PC access while on the road, without involving the entire IT department in the process.

Home RF Gateways and Routers

The following section discusses wireless gateways that provide wireless access on the Home RF standard.

Motorola BB160 Ethernet Gateway

All Motorola Wireless Home Gateway products use the HomeRF standard. One example is the Motorola BB160 Ethernet Gateway. Like most home wireless access points, the BB160 Ethernet Gateway provides networking of all wireless-enabled computers and sharing of an existing Internet connection. Only a single IP address is exposed to the Internet. The Motorola BB160 extends a network to multiple rooms and floors, sharing access with up to ten computers within a span of 150 feet.

Cayman 3200 HW

The Cayman Systems' Cayman 3200 HW is an ADSL gateway (hub/router) fully integrated with the Home RF specification. The Cayman 3200 HW allows sharing of wired Ethernet connections, DSL, and other high-speed lines, as well as wireless access for all enabled devices within 150 feet. The Cayman 3200 HW uses NAT security (see Chapter 5); thus, only one IP address is exposed.

iPAQ Connection Point

Compaq's iPAQ Connection Point (see Figure 6-12) provides home connectivity of all types, enabling wireless access via Home RF, wired connections via Ethernet and 56K dial-up, and firewall-protected Internet access to all connected devices. This broad onramp device allows home users with an existing Ethernet to move into the Home RF arena, leaving their existing network intact.

Figure 6-12: The iPAQ Connection Point (photo courtesy of Compaq Computer Corporation)

The Wireless Home

The wireless home hoopla gave way to the realization that consumers were really not that interested in kitchen devices or toiletries that worked liked computers. Still, there is an active interest in using wireless devices to make homes more secure, monitor rooms at a distance, and automate environmental factors, such as temperature, air humidity, lighting, and multimedia.

SmartHome provides access to many home wireless devices, such as the X10 Activehome Automation Kit. You can control lighting and electronics of all types and schedule programmable events that power on and off electronic devices from your PC. X10 is a home automation firm that provides wireless access to all kinds of household devices. You can install and remotely control motion sensors, remote door and window access, voice activated controls, and all kinds of video surveillance paraphernalia.

Summary

Obviously, there are many, many more devices we could discuss. This chapter merely scratches the surface. Because mobile wireless technology as we know it today is fairly new, it is fascinating to discover how various manufacturers devise ways to keep people connected wherever they may be.

This concludes the product-oriented segment of the book. For the remainder, technology applications, specifications, security, and the wireless Web will be discussed. In the next chapter, first generation (1G) techs and specs will be covered.

Part II

Techs and Specs

Chapter 7

1G Techs and Specs

In This Chapter

First-generation wireless technologies, commonly called *1G technologies,* have been deployed worldwide since the 1980s. Understanding these old analog voice-only technologies, and the cellular radio networks they work with, is an important step toward understanding the newer digital technologies that most people think of when they talk about wireless devices. Not only are the 1G technologies the forerunners of the cooler second-generation (2G) and later technologies, but in many cases, the newer technologies and protocols coexist and even run on top of the 1G infrastructure. The time you spend reading this chapter will pay dividends in a stronger foundation for understanding the latest wireless devices and communication systems. In general, the following topics are covered:

✓ Understanding cellular basics

✓ Understanding first-generation (1G) wireless technology

✓ AMPS

✓ NMT

✓ TACS/ETACS

Understanding Cellular Basics

To thoroughly understand wireless devices, you must understand the basics of cellular telephone systems. Most current wireless devices — no matter how sophisticated the technology they use to give you high-speed, always-on, digital connectivity to the universe — depend on the existing cellular phone systems that were deployed in the 1980s and 1990s. It is definitely worth your while to understand the basic functioning of these systems.

Cells

The basic unit of any cellular system is a *cell.* A cell is a geographic area in which there is radio coverage by a particular cellular phone system. The size of any given cell within a system varies, depending on the geography of the cell and the number of mobile devices that each cell is expected to serve.

Geography is important to cell size because a wireless device must be able to send signals to, and receive them from, the antennas that connect the device to the cellular system. The signals used in cellular systems are low-powered and affected by the terrain in the cell. In general, cells in rough terrain need to be smaller than cells in open plains. Likewise, cells that contain lots of buildings or other man-made obstacles, such as power lines, need to be smaller than those that do not.

The number of wireless devices a cell must support is important because the number of simultaneous wireless *channels* available within a cell is limited (for more information about this topic, see the "Frequencies and Channels" section later in this chapter). The greater the number of wireless devices a cell supports, the higher the odds that all the channels will be in use at the same time, and the greater the likelihood that devices will not be able to connect to the cellular system when they need to. Thus, cells in cities tend to be smaller than cells in rural areas, as the higher population density in the city will likely increase the number of supported wireless devices in any given area. Cells are typically two to ten miles across, although in high-density urban areas, cells could be as small as half a mile in diameter.

Cells can also be subdivided, or split. *Cell splitting* occurs when a cell can no longer support the number of wireless devices within it. The service provider that maintains the cellular system can turn one large cell into multiple smaller cells that cover the same area. More cells within the same geographic area means more available channels within that area, thereby allowing the system to support more users.

So far, we've talked a lot about cells and the factors that affect the size of cells, without touching on the issue of how cell boundaries are determined and how radio signals get into and out of them. To understand those issues, you need to know about cell sites and base transceiver stations.

CELL SITES AND BASE TRANSCEIVER STATIONS

A *cell site* is the point from which signals are broadcast into a cell. Located at the cell site is a *Base Transceiver Station* (*BTS*), the radio station and antennas that communicate with wireless devices within a cell. You can easily spot cell sites in your area by looking for those ugly cell phone towers (the BTS located at the cell site) that seem to have popped up all over the place in the last 15 years.

When most people think of cell sites (if they think about them at all), they envision the cell site (and the hardware that is located there) at the center of the cell, transmitting outwards, with the boundary of the cell forming a radius, as shown in Figure 7-1. They are wrong.

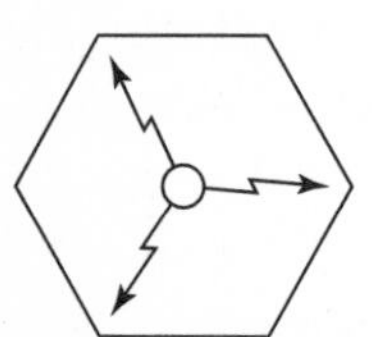

Figure 7-1: The common view of the relationship between a cell site and the cell it serves

In reality, cell sites are located on the edges of cells. Each cell site typically serves three cells, transmitting from separate smaller antennas into each cell. The shape of cells formed by such a layout is best considered a hexagon, with three of the six vertices located at cell sites, as shown in Figure 7-2.

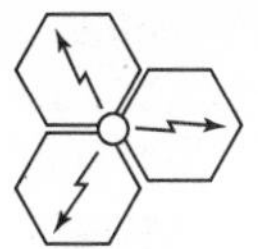

Figure 7-2: The actual relationship between cell sites and the cells they serve

In this common configuration, three cell sites and their associated BTS serve each cell. The particular BTS that communicates with a particular wireless device is determined by the strength of the radio signal from the wireless device, as measured at each BTS. Likewise, the particular cell a wireless device communicates with is determined by signal strength. The key to making all this work smoothly is the Mobile Telephone Switching Office (MTSO).

MOBILE TELEPHONE SWITCHING OFFICES

A *Mobile Telephone Switching Office* (*MTSO*) is a facility that connects to all the BTSs, providing the computers that control the switching between base stations, as well as the connection between the cellular system and the *Public Switched Telephone Network* (the old-fashioned wired phone system, or *PSTN*). In addition, the computers at the MTSO monitor call status, authenticate devices trying to communicate with the cellular system, and manage the *handoffs* that occur when wireless devices move between cells. But before we discuss these topics, you need to learn about cellular system frequencies and channels.

Frequencies and Channels

The radio transmitters and receivers in every cellular system operate within an assigned frequency range. The AMPS system (Advanced Mobile Phone System), for example, works within a frequency range of 824 to 894 MHz. To ensure that communication takes place, a certain amount of that frequency range must be used to carry the information that moves between the wireless device and the BTS. However, there are many frequencies between 824 and 994 MHz, and many ways to divide up that range. Somehow, the BTS and the wireless device must choose a frequency range to use, and they need to choose one that isn't being used by another wireless device in the same cell. This is where the concept of a *channel* comes in handy.

In the United States, the Federal Communications Commission (FCC) is responsible for allocating the use of the radio-frequency spectrum. Since the FCC is a government agency, it is useful to note that the process of allocating the radio spectrum can vary, depending on the political climate in Washington, D.C. Such concerns are not too important for 1G systems, most of which have been in place for more than a decade. But such concerns are a very big deal when it comes to current and future allocations. Who gets the spectrum, and how much they pay for it, will partly determine the wireless services we have in the future.

A channel is a predefined range of frequencies used to carry a single connection between a wireless device and a BTS. Returning to AMPS for a concrete example, each AMPS channel is a 25 kHz, non-overlapping band of frequencies between 800 and 900 MHz. Dividing the available radio spectrum into channels distinguished by frequency is known as *Frequency Division Multiple Access* (*FDMA*).

Whenever a wireless device connects to a cellular network that uses channels, that device is assigned to an unused channel at the time it connects to the system. Within this channel, there should be no interference from anything else within the cellular system.

Channels are a great way to manage communications within a cellular system, but there can only be a finite number of channels within any range of frequencies. Once all the frequencies in a particular region of the cellular system are in use, the system cannot support any more connections in that area. The ability to deploy more cells, and thus more channels, is one reason why systems with small cells are favored over systems with large cells.

To avoid interfering with neighboring cells, groups of adjacent cells, known as *cell clusters*, use different channels from those used in the other cells of the cluster. Figure 7-3 shows the available channels divided among seven cells.

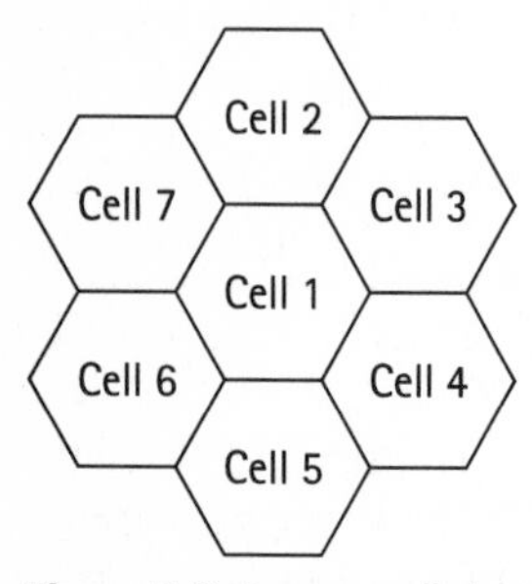

Figure 7-3: A seven-cell cluster using seven nonoverlapping sets of channels to avoid interference between cells

In Figure 7-3, the set number within each cell represents one of seven sets of nonoverlapping channels. Again, using AMPS as an example, if 409 channels are available for general use in the cellular system and those channels are divided into seven equal sets, you end up with 58 channels per set.

Why 409 channels? See the "AMPS" section later in this chapter for all the gory details about dividing the available radio frequency spectrum.

Dividing the channels this way enables each of the seven cells to support as many channels as possible, without using any of the same channels as its neighbors, thereby eliminating the possibility of interference between wireless devices in adjacent cells because they are all talking on different channels. But avoiding channel duplication in nearby cells also adds a lot of complexity to the system, while reducing the number of devices each cell can support (since each cell can use fewer channels).

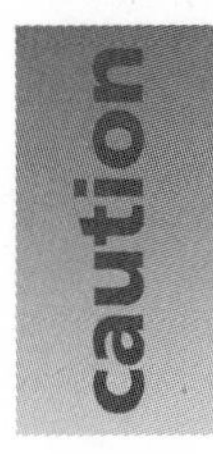

When talking about cellular systems, it makes the most sense to refer to the cells as if they were hexagonal areas, butted up against each other to ensure 100 percent coverage of any geographical area without overlap. In reality, though, a cellular system is composed of radio transmitters spraying signals in all directions. These signals don't magically stop at the edge of a cell, or provide the same signal level across an entire cell. Avoiding the use of the same channels in adjacent cells is one way to prevent these messy realities from mucking up the system.

Nevertheless, reducing the number of channels each cell can use and dividing the available channels in the way described here is actually the key to vastly increasing the capacity of the overall cellular system. Dividing the channels in this way enables huge numbers of wireless devices to talk on the same channel simultaneously without interfering with each other. The key to this activity is frequency reuse.

FREQUENCY REUSE

Frequency reuse is a technique for reusing the available radio frequencies (and hence the channels those frequencies are divided into) in noncontiguous cells. No two adjacent cells can interfere with each other, because they don't use any of the same channels.

Eliminating channel interference is helpful, but the cellular system is still only using each channel once. Conceptually, you could dispense with six of the seven cells in the figure by putting one massive BST in the center of the area, and letting it use all of the available channels. That would certainly reduce the complexity of the system.

But suppose you reuse frequencies (and the associated channels) in noncontiguous cells. Now you can increase the total number of wireless devices your cellular system supports by using each channel more than once. To really visualize this, look at Figure 7-4, which shows the cell cluster from Figure 7-3 and three adjacent cell clusters.

If you examine Figure 7-4 closely, each set of channels is reused in each cell cluster, and no set of channels is used in two adjacent cells. This pattern can continue indefinitely, with each channel available in each cell cluster, and no possibility of interference by an adjacent cell using the same channel. In effect, frequency reuse allows a limited number of channels to support a vastly greater number of wireless devices, limited only by the number of cells in the cellular system. Add more cells, support more devices.

This also explains why small is good when it comes to cell size. The smaller the cell, the more there are. And the greater the number of cells, the greater the opportunity to reuse frequencies, multiply the effective number of channels in the system, and maximize the number of wireless devices the system can support in a given area.

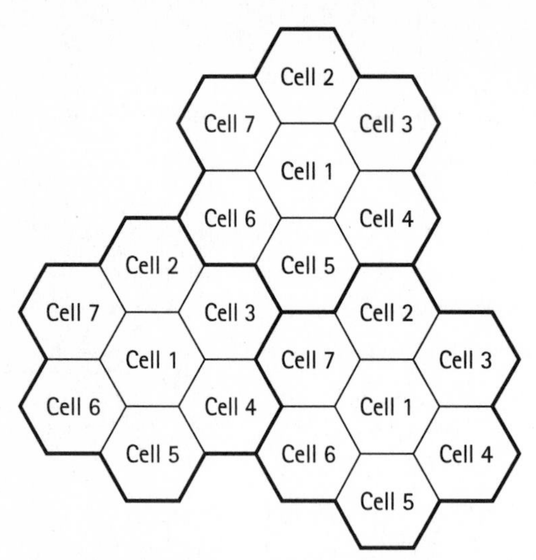

Figure 7-4: Multiple adjacent seven-cell clusters, showing how, by dividing channels among cells properly, each channel (or set of channels) can be repeatedly reused

So the cellular model for a wireless communication system solves the problem of supporting a large number of wireless devices. There is, however, one additional problem to consider. Wireless devices have an annoying tendency to move, which isn't a big deal as long as the device stays inside one cell. But eventually, some wireless device somewhere is going to make things difficult and move from one cell to the next. The cellular system needs a way to deal with this smoothly. The MTSO manages this exercise, known as *handoff.*

HANDOFF

Proper handoffs are vitally important to a cellular system. It wouldn't do at all for wireless devices to lose their connections when they move between cells. Imagine holding a mobile-phone conversation with your boss while you are driving down the highway. Now imagine that every 10 minutes or so, you lose the connection and have to call back your boss . That's what cellular-system use would be like if there were no mechanism for handing off a wireless device from one cell to another.

A handoff is the process by which the wireless device, under the direction of the MTSO, automatically switches from its current channel to a stronger channel in another cell. Here's a simplified description of how a handoff works:

1. As the wireless device moves, the signal for the channel it is using starts to fade.
2. The BTS communicating with the wireless device detects that the signal is fading and requests a handoff.
3. The MTSO determines which BTS has the strongest signal from the wireless device. Since the wireless device is moving across a cell boundary, the BTS with the strongest signal is one in a different cell, which means it is using a different set of channels.

4. The MTSO assigns the wireless device to the new BTS and specifies a new channel for the device to use.

5. The wireless device switches channels and begins to communicate with the new BTS in the new cell.

Assuming that all goes smoothly, no noticeable break occurs in the connection between the wireless device and the cellular system, which resolves the last major issue in implementing a cellular system.

Cellular Basics Wrap-Up

A cellular radio system is one of the more complex systems you will face in your day-to-day life. The preceding pages addressed the main features of a typical cellular system, without going into the low-level details. Here, again, are the main points:

- ✓ The cellular phone systems deployed in the 1980s and 1990s underlie much of the ultramodern, continuously connected digital wireless technology everyone is so excited about today.
- ✓ Cells are the geographic regions served by one or more Base Transceiver Stations, each located at a cell site.
- ✓ The geography of a cell and the number of wireless devices it must support affect the size of the cell.
- ✓ The band of radio frequencies allocated to a cellular system is divided into channels.
- ✓ Each cell can support only a limited number of wireless devices, as each BTS can serve a limited number of channels.
- ✓ Cells can be split into smaller cells when necessary to serve an increased number of wireless devices.
- ✓ The cells within a cell cluster all use a different set of channels, so no cell in the cluster can interfere with any other.
- ✓ Frequency reuse is a technique where, through careful division of sets of channels within cell clusters, the same channel can be used in multiple cell clusters without risk of interference.
- ✓ Within the limits of practicality, the smaller the cell size the better. Smaller cells mean more cells, which means more reuse of channels and an increase in the number of wireless devices the overall cell system can support.
- ✓ As they move about within the area covered by the cellular system, wireless devices are automatically handed off from BTS to BTS and cell to cell, all under the control of the MTSO.

on the web

If you are really hungry for detailed information on the workings of cellular systems in general, and the analog cellular phone system deployed across much of the United States in particular, take a look at the Cellular Telephone Basics page at TelecomWriting.com. Tom Farley's site contains plenty of details about thetechnology, and even more about the history of telecommunications, including personal anecdotes and writings by people who actually built the system. You can access Farley's site at `www.telecomwriting.com/Cellbasics/Cellbasics.html`.

Understanding First-Generation (1G) Wireless Technology

First-generation wireless systems, referred to as *1G* systems, came into general use during the early and mid-1980s. Understanding the technologies that 1G systems employ and the strengths and weaknesses of these technologies will provide you with a basic foundation for understanding the evolution of wireless standards and wireless devices.

Key Characteristics of 1G Systems

Table 7-1 lists the two key characteristics of 1G systems.

TABLE 7-1 **Key 1G System Characteristics**

Characteristic	Technology
Modulation	FM (analog)
Switching	Circuit-switched

MODULATION

A key 1G system characteristic is the technology used to modulate a signal. To transmit a signal wirelessly, you need to convert it into electromagnetic waves — more specifically, radio waves (for the kinds of systems we are interested in). These radio waves have two parts: a carrier wave and a signal.

A *carrier wave* is a continuous radio wave with known characteristics. Because both the transmitter and the receiver know the characteristics of the carrier wave, a *signal* can be represented as a change to the characteristics of the carrier wave. The transmitter can modify the carrier wave in accordance with the signal, and the receiver can detect those modifications and reconstruct the original signal.

In the case of FM modulation, the transmitter encodes the signal on the carrier wave by varying the frequency of the carrier wave in proportion to the level of the input signal. FM modulation is an analog technology, which means that the signal being modulated onto the carrier wave varies continuously, as opposed to the digital method in which only a few values can be specified.

1G systems were designed to carry voice signals, which are analog. The systems eventually connected to the *Public Switched Telephone Network* (*PSTN*), the worldwide phone system we grew up with. The PSTN is (or was, at the time) an analog system. And wireless digital devices of the kind covered in the first part of this book were scarcely a gleam in a designer's eye.

Thus, building 1G systems as analog systems made perfect sense — for awhile.

SWITCHING

Switching is the other key characteristic of 1G systems. A *circuit-switched* system is one in which a dedicated connection is set up between the transmitter and receiver and maintained for the duration of the communication. In a 1G system, one channel (out of the finite number available within any area of the wireless network) is dedicated to a wireless device for the entire time the wireless device is connected to the wireless network.

As with the use of analog signals, this characteristic makes perfect sense, given the context within which 1G systems were developed. The PSTN is (or was) circuit-switched, so designing 1G systems as circuit-switched systems just makes them more compatible with the regular telephone system they are connected to.

Unfortunately, the capacity of circuit-switched systems is limited to the number of channels they support in any given area. This issue was initially addressed by the development of the cellular system, but circuit switching has more serious ramifications when wireless devices are using the network.

The Ramifications of the Key 1G Technologies

The key technologies used in 1G systems, FM modulation and circuit switching, are appropriate for wireless systems used primarily to carry voice signals. But FM modulation and circuit switching are much less desirable when a wireless system needs to carry digital data in addition to, or in lieu of, voice data. There are three main reasons for this:

- ✓ **Circuit switching leads to connect-time billing.** Since a connection on a circuit-switched system ties up one channel for the duration of the connection, users get billed by the minute of connect time. This works fine for voice services, as users are only connected to the wireless system when they are actually conducting a call. Digital wireless devices work best when they have a continuous connection to the wireless system; thus billing by the minute would be obscenely expensive.
- ✓ **Analog encoding is an inefficient way to transmit digital data.** Most of the wireless devices we are concerned with are digital devices. Converting signals from digital form in the device to analog form to be modulated onto the signal, demodulated at the other end, and finally converted from analog to digital, is clumsy and inefficient.
- ✓ **1G systems are slow at transferring data.** Partly due to all the needed conversions between analog and digital, and partly because it doesn't take a lot of bandwidth to transmit voice-quality analog signals, 1G systems can only transfer data at a low rate.

Therefore, 1G systems, while a great advance for their time — and still quite useful for certain types of wireless telecommunication (voice) — are at best marginally suitable for use with data devices. This situation drives the need for the 2G and higher systems that have appeared. Even so,

1G systems are still widespread and actually *do* have a part to play in the development and deployment of 2G and higher systems.

AMPS

AMPS, the *Advanced Mobile Phone System*, is one of the original cellular-phone air-interface standards deployed throughout much of the world. Even now, almost 20 years after its initial deployment, AMPS is in widespread use, particularly in the United States. Table 7-2 shows the main characteristics of AMPS.

TABLE 7-2 AMPS at a Glance

Characteristic	Value
Developer	AT&T Bell Labs
Initial Deployment	1983 (trials began in the 1970s)
Wireless Generation	1G
Signal Type	Analog, Frequency Modulated (FM)
Spectrum Division	FDMA
Frequency	824–894 MHz
Channel Spacing	30 kHz
Switching	Circuit-switched

For the heavy-duty details about AMPS, you can go to the source and buy a copy of TIA/EIA/IS-91A, or a later revision. A copy of this standard is available from the Telecommunications Industry Association (TIA) at www.tiaonline.org. At the time of this writing, the standard is available in either hard copy or electronic form for $265.

AMPS was deployed by Ameritech and declared part of the U.S. telecommunications standard by the FCC. While direct government involvement in setting the standard is not typical procedure today, it did allow a nationwide cellular telephone network to develop quickly, without the confusion and uncertainty that competition would have caused.

AMPS has been quite successful, spreading from the United States to Canada andMexico, and to much of the rest of Central and South America. AMPS also became the basis for TACS and ETACS, two other 1G systems.

You can find out more about TACS and ETACS in the section "Understanding TACS," later in this chapter.

Some Important Aspects of AMPS

A detailed analysis of AMPS is beyond the scope of this book. However, several aspects of AMPS bear closer examination: channels and bands, special numbers, and roaming and security concerns.

CHANNELS AND BANDS

AMPS operates at frequencies between 824 MHz and 894 MHz. This spectrum is divided into 1,664 channels, spaced 30 kHz apart. The signal that is modulated onto each of the channels used for voice communication is a 3 kHz analog signal, the same type of signal that is carried by the PSTN.

Because of FCC concerns about the possibility of creating monopolies on mobile phone service in any given area, the set of possible channels is divided into two bands: the *A Band*, and the *B Band*. The A Band is reserved for use by a non–wire line carrier (a mobile phone company), while the B Band is reserved for use by a wire line carrier (the local phone company).

Dividing 1,664 channels evenly between two bands yields 832 channels per band. These channels are then grouped into 416 *channel pairs*, with 45 MHz between the two channels in the pair. A wireless device transmits on the lower-frequency channel in the pair and receives on the higher-frequency channel. Table 7-3 shows the frequencies used by each band.

TABLE 7-3 AMPS A and B Band Frequencies

Band	Frequency Range	Used for
A	824 MHz to 835 MHz	Transmitting from mobile device to base station
A	845 MHz to 846.5 MHz*	Transmitting from mobile device to base station
B	835 MHz to 845 MHz	Transmitting from mobile device to base station
B	846.5 MHz to 849 MHz*	Transmitting from mobile device to base station
A	869 MHz to 880 MHz	Transmitting from base station to mobile device
A	890 MHz to 891.5 MHz*	Transmitting from base station to mobile device
B	880 MHz to 890 MHz	Transmitting from base station to mobile device
B	891.5 MHz to 894 MHz*	Transmitting from base station to mobile device

** These frequencies are part of an additional 5 MHz of bandwidth that was allocated for AMPS use in 1986. This gave each operator an additional 83 channel pairs to use.*

Finally, anywhere from 7 to 21 of the channels are *control channels*, where digital data can be transmitted at 10 Kbps and used to manage the wireless system. The system is left with anywhere from 395 to 409 distinct voice channels. So in a best-case scenario, any given AMPS cell can support 409 simultaneous voice conversations. Of course, that number doesn't take into account that a cell will probably only be able to use 1/7th of the available channels so that frequency reuse is possible. Now each cell is down to 58 simultaneous conversations. It is no wonder the towers are so close together in urban areas. They need to be able to support a significant number of users.

SPECIAL NUMBERS

Every wireless device that uses AMPS must be programmed with several special numbers. These numbers serve, among other things, to uniquely identify the device or the cellular system the device belongs to. The following list describes these five special numbers:

- ✓ **Electronic Serial Number (ESN):** This 32-bit number uniquely identifies a particular wireless device. The ESN is burned into the device's memory when it is manufactured.
- ✓ **Home System Identification Number (SID or SIDH):** This 15-bit number identifies the *wireless service provider* (*WSP*) a particular device belongs to. In cases where the WSP distinguishes between local area and other areas within the system, the SID can vary by location.even though the WSP is the same. The SID is entered using the device's keypad, and can be changed the same way if necessary.
- ✓ **Mobile Identification Number (MIN):** This 34-bit number is the binary representation of the wireless device's 10-digit telephone number. The MIN is entered using the device's keypad, and can be changed the same way if necessary.
- ✓ **First Paging Channel:** This 11-bit number identifies the first channel the wireless device should scan when monitoring the control channels looking for commands from the wireless network. Normally, the wireless device will receive commands on this channel that direct it to monitor a different channel. This reduces the load on the first paging channel. The first paging channel is entered using the device's keypad, and can be changed the same way if necessary.
- ✓ **Last Paging Channel:** This 11-bit number identifies the highest channel number the wireless device should scan when monitoring the control channels looking for commands from the wireless network. The last paging channel is entered using the device's keypad, and can be changed the same way if necessary.

ROAMING

One of the benefits of using AMPS is that there is complete – more or less – coverage of the United States. In other words, if you go almost anywhere in the United States, you can turn on your AMPS mobile phone and place a call.

The United States is not covered by a single WSP. Instead, the AMPS cellular network your phone connects to could be owned and operated by one of many WSPs. The technology is the same, but the owner is different. And that owner wants to get paid when you connect to his cellular network. It is at this point that the SID described in the preceding section comes into play. Your mobile phone transmits the SID when it connects to the wireless network. With the SID, any network can determine the identity of your home network, and pass along the information necessary for you to get billed. (Now you know where those roaming charges on your mobile phone bill come from.)

Roaming can also come into play if your home network bills you differently when you are outside your local calling area, even though you are still within the home network.

SECURITY CONCERNS

One last consideration when dealing with AMPS is the standard's lack of security. It isn't very hard to eavesdrop on AMPS phone calls. All it takes is a simple scanner such as those used to monitor police-band radio traffic. Since AMPS uses analog signals, it is hard to encrypt the signal, the way you could if it were digital. If wireless security is one of your primary concerns, think long and hard before using AMPS to carry any sensitive information. A later-generation technology will likely suit you better.

You'll find an overview of wireless device security in Chapter 14.

Making a Call

At this point, you should know enough about AMPS to follow the progress of a phone call that originates at a mobile phone, as well as one that originates in the PSTN. Tables 7-4 and 7-5 cover the basic sequences, without going into minute detail. The "access channel" and "setup channel" referred to in the tables are two additional control channels.

The following sequence of steps is adapted from "Cellular Telephony II," by Brian Oblivion, and "Cellular Telephone Basics: AMPS & Beyond," by Tom Farley. You can find Oblivion's document at `www.aftermath.net/cellular/cell/texts/cell2-rdt.txt`, and Farley's at `www.telecomwriting.com/Cellbasics/Cellbasics.html`.

A CALL ORIGINATING AT A MOBILE PHONE

Table 7-4 describes the sequence of steps that occur when a call originates at a mobile phone.

TABLE 7-4 AMPS Processing of a Phone Call Originating at a Mobile Device

Step	MTSO	BTS	Mobile Phone
1		Continuously transmits control data, including paging information, on the first paging channel.	
2			User turns on the mobile phone, which scans and locks on to the paging channel.
3			User initiates a call.
4			Scans, then locks on to the access channel.
5			Takes control of the setup channel and sends a service request.
6		Reformats the service request and determines the direction of the mobile phone before passing the service request to the MTSO.	
7	Selects a voice channel and sends it to the BTS.		
8		Reformats the channel designation message from the MTSO and sends it to the mobile phone using the access channel.	
9			Tunes to the designated voice channel, then receives and retransmits (transponds) a *Supervisory Audio Tone* (*SAT*).
10		Detects the SAT and notifies the MTSO that the user has gone *off-hook* (picked up the phone).	
11	Receives the off-hook signal from the BTS and completes the call through the PSTN.		

A CALL ORIGINATING ON THE PSTN

Table 7-5 describes the sequence of steps that occur when a call originates on the PSTN.

TABLE 7-5 **AMPS Processing of a Phone Call Originating on the PSTN**

Step	MTSO	BTS	Mobile Phone
1		Continuously transmits control data, including paging information, on the first paging channel.	
2			User turns on the mobile phone, which scans and locks on to the paging channel.
3	Receives incoming call from the PSTN, translates it as necessary, and sends a paging message.		
4		Reformats paging message for mobile phone and transmits it over the paging channel.	
5			Detects paging message, then scans and locks on to the access channel.
6			Takes control of the setup channel and sends a service request.
7		Reformats the service request and determines the direction of the mobile phone before passing the service request to the MTSO.	
8	Selects a voice channel and sends it to the BTS.		
9		Reformats the channel designation message from the MTSO and sends it to the mobile phone using the access channel.	

Continued

Table 7-5 **AMPS Processing of a Phone Call Originating on the PSTN** *(Continued)*

Step	MTSO	BTS	Mobile Phone
10			Tunes to the designated voice channel, then receives and retransmits (transponds) a *Supervisory Audio Tone* (*SAT*).
11		Detects the SAT and notifies the MTSO that the user has gone *off-hook* (picked up the phone).	
12	Receives the off-hook signal from the BTS and sends an alert order.		
13		Reformats the alert order and sends it to the mobile phone on the voice channel using *blank-and-burst*.	
14			Rings to alert the user to an incoming call, then sends a tone to signify that the phone is *on-hook* (phone has not been picked up).
15		Detects tone and sends on-hook signal.	
16	Detects on-hook signal and provides a ring to the caller through the PSTN.		
17			Terminates the on-hook tone when user answers the call.
18		Detects absence of on-hook tone and transmits an off-hook signal to the MTSO.	
19	Receives off-hook signal from the BTS, terminates audible ring, and completes the call through the PSTN.		

NMT

NMT, the *Nordic Mobile Telephony system*, was developed by Ericsson and Nokia and deployed in 1981. NMT became the 1G standard cellular phone system in Denmark, Finland, Norway, and Sweden, and has been deployed in more than 40 countries around the world, including Russia, the Middle East, and parts of Asia.

There are two versions of NMT. The original version is now known as NMT 450, short for NMT operating on a frequency of 450 MHz. A newer version of NMT, NMT 900, operates on a frequency of 900 MHz.

NMT is noteworthy for several reasons. It was the world's first multinational cellular network, allowing roaming between countries. It was specified jointly by the four Nordic countries, with these roaming capabilities in mind. Although we cannot confirm this through first-hand experience, we have seen it reported in numerous places that NMT provides high-quality voice signals and functions well in the rough terrain of Scandinavia. And NMT, at least NMT 450, is still widely used, particularly in Russia and non-Western European countries.

Table 7-6 shows the main characteristics of NMT. Note that the developers continue to upgrade NMT. For example, NMT 450i is an improved version of NMT, drawing from NMT 450 and NMT 900. Consider Table 7-6 to be a snapshot of this 1G technology.

TABLE 7-6 **NMT at a Glance**

Characteristic	Value
Developer	Ericsson & Nokia
Initial Deployment	NMT-450, 1981
	NMT-900, 1986
Wireless Generation	1G
Signal Type	Analog, Frequency Modulated (FM)
Spectrum Division	FDMA
Frequency	NMT-450, 453–468 MHz
	NMT-900, 890–960 MHz
Channel Spacing	NMT-450, 25 kHz
	NMT-900, 12.5 kHz
Switching	Circuit-switched

In April 2000, Lucent Technologies and Radio Design AB of Sweden announced plans to work together on migrating the NMT system to CDMA 450, a 3G variant on the CDMA standard. See the section "Understanding 3G Concepts" in Chapter 10 for more on CDMA 450.

TACS/ETACS

TACS, the *Total Access Communication System*, was developed in the 1980s by Vodafone. Two major versions of this technology are in use: TACS and ETACS.

TACS

TACS is basically a variant on AMPS, operating in the 900-MHz range instead of the 800-MHz range. TACS also uses 25-MHz channel spacing, instead of the 30-MHz spacing used by AMPS. Table 7-7 shows the main characteristics of TACS.

TABLE 7-7 TACS at a Glance

Characteristic	Value
Developer	Vodaphone
Initial Deployment	United Kingdom, 1985
Wireless Generation	1G
Signal Type	Analog, Frequency Modulated (FM)
Spectrum Division	FDMA
Frequency	890–950 MHz
Channel Spacing	25 kHz
Switching	Circuit-switched

In 1979, the World Administrative Radio Conference specified 1,000 channels for mobile phone use. These channels were located between 890 MHz and 950 MHz, and spaced 25 kHz apart. In 1995, TACS service began, with Vodafone and Cellnet set up as the competing A- and B-band service providers. Each received 220 channels, a number that was quickly boosted to 300 channels.

TACS systems remain in use today, but they are gradually being superseded by GSM.

ETACS

ETACS, the *Extended Total Access Communication System*, came into existence to resolve a specific problem with TACS. Table 7-8 shows the main characteristics of ETACS.

TABLE 7-8 **ETACS at a Glance**

Characteristic	Value
Developer	Vodafone
Initial Deployment	United Kingdom, 1987
Wireless Generation	1G
Signal Type	Analog, Frequency Modulated (FM)
Spectrum Division	FDMA
Frequency	872–960 MHz
Channel Spacing	25 kHz
Switching	Circuit-switched

Within a few years of the deployment of TACS, the number of users in the Greater London area had reacheda level that was threatening to overwhelm the local cellular capacity. To address that issue, TACS was extended by adding an additional 640 channels, 320 for each band. These channels occupied bandwidth reclaimed from military use, at 872–880 MHz and 917–933 MHz. These additional channels are known as ETACS.

Summary

Most people don't think about the basics of cellular radio networks when they choose a wireless device. At first blush, there may seem little connection between 1G cellphone communication technologies and the "promise" of wireless via technologies that are just now developing. But a thorough understanding of concepts is important for wireless-device users. Because most current digital wireless technologies are built on top of the existing analog cellular radio network and its 1G wireless technology, the characteristics of these systems greatly influence the performance of most 2G and later technologies, which we discuss in the next three chapters.

Chapter 8

2G Techs and Specs

In This Chapter

Second-generation (2G) wireless technologies are the descendants of the original analog 1G technologies initially deployed in the 1980s. They are designed to address some of the short-comings of 1G — in particular, the limited number of users those systems could support, their often-poor voice quality, and their very limited ability to carry digital data. In this chapter, 2G wireless technologies will be discussed.

- ✓ Understanding 2G concepts
- ✓ CDMA
- ✓ CDPD
- ✓ D-AMPS
- ✓ GSM
- ✓ PCS
- ✓ TDMA

The Current State of 2G Technologies

Companies developed a number of different solutions to resolve these problems. CDPD (Cellular Digital Packet Data), for example, tried to ensure the maximum reuse of the existing cellular infrastructure. In effect, CDPD layers a digital communication protocol onto the old 1G AMPS technology.

Other providers decided to develop new technologies, such as CDMA (Code Division Multiple Access). Such systems employ some of the same concepts as 1G systems, but function very differently, requiring significant investments in new infrastructure.

As of this writing, 2G technology is state-of-the-art for the United States. 2G technologies are deployed and widely available in most regions of the country, although the coverage for each technology varies.

Even when you have coverage, the specific capabilities you have depends on whom you buy your wireless service from. Verizon Wireless provides different capabilities than Sprint PCS — and both differ from VoiceStream. Devices that work on one network probably won't work on another. Thus, although decent technology is available in the U.S., understanding and using 2G technology is confusing, due to incompatible networks, incomplete coverage, and inconsistent capabilities.

Connecting from My House

One of us knows from personal experience that coverage varies greatly depending on location and technology. This case study details his experience.

My mobile phone (a Motorola V8160, if you must know) works great. Most of the time, I have a nice, solid digital signal that allows me to send and read e-mail, check my stocks, and so on. I also have a Palm Vx with a wireless modem that works just as well around town. In other words, when I'm out and about, I can read e-mail, check my stocks, and so on with either device. However, being at home presents some problems.

At home, I get the same great performance with my Palm and wireless modem combination that I have when I'm around town. But I have absolutely no digital connection with my phone. All it allows me to do is make phone calls. This performance difference results from the technologies that the two devices use.

The wireless modem I use with the Palm Vx (a Novatel Minstrel V) is a CDPD modem. The Motorola phone is a dual-mode phone that uses the Verizon Wireless CDMA network for its digital communications.

My house is at the foot of a small mountain, which separates me from the rest of civilization. Although the wireless modem can connect with the nearest CDPD-equipped transceiver, the phone simply cannot make a connection to any of the nearby CDMA-equipped transceivers.

So beware if you go shopping for a wireless device. If you live or work near the edge of a coverage area (the vendors you are working with should be able to provide coverage maps for your area), try to get a device on a trial basis so you can test it in places where you'll actually be using it. If all else fails, find out what device is being used by the people who live or work in those locations.

In Europe and much of the world, the situation is different. GSM is standard nearly everywhere (there are over 500 million GSM users worldwide). Coverage is approaching 100 percent across Europe, and is available in Africa, portions of Asia, and ever-growing segments of the U.S. The services provided on the various GSM networks are pretty standard, and users can move from one GSM network to the next with little trouble. So, while technologists argue the relative merits of 2G technologies, the U.S. flounders and falls ever farther behind the rest of the world.

Understanding 2G Concepts

Aside from the technology-specific concepts covered later in this chapter, most of the 2G-technology concepts are the same as those you learned in our discussion of 1G technologies in Chapter 7. (The 1G terms are important because 2G technologies employ many of the same concepts as 1G.) A basic understanding of cells, channels, frequency and frequency reuse, and handoff is particularly important.

At this point, the only concept that needs to be introduced is that of *digital encoding*. 1G technologies use *analog encoding* to carry voice messages. Analog-encoded signals are made up of continuously varying information. Because the human voice — and all the other sounds humans make — are analog in nature, it seems logical to use analog encoding to carry those sounds through a wireless network.

Digital encoding, on the other hand, takes the analog sounds we produce, measures their levels thousands of time per second, and converts those levels into binary numbers. It then transmits the binary numbers across the wireless network. Once they reach their destination, the signals are converted from digital to analog form, which can be made suitable for our ears.

Although converting voice signals to digital form and back again seems like a lot of unnecessary work, it actually confers several benefits. These benefits are discussed in more detail in the sections on the individual technologies, as they make different uses of the digital signal.

Digital encoding makes a lot of sense for wireless devices that work with digital data as well as sound. Moving digital data on a digital wireless network makes perfect sense and allows for faster, more accurate and more efficient data transfers than on 1G networks.

As you'll learn in the following sections, the advantages of digital encoding are such that all the 2G technologies are digitally encoded.

Code Division Multiple Access

Code Division Multiple Access, or CDMA, is viewed by many people as the core technology for the future of wireless devices. Taking advantage of some very sophisticated (and patented) mathematical techniques, CDMA provides many advantages over FDMA (Frequency Division Multiple Access) and TDMA (Time Division Multiple Access) technology, such as

- ✓ More efficient use of bandwidth
- ✓ Improved noise and interference resistance
- ✓ Simplified system planning
- ✓ Fewer dropped calls
- ✓ High security

In July 2001, Nokia, then the largest maker of cell phones in the world, expanded its CDMA licensing agreement with Qualcomm to include the production and sale of CDMA infrastructure equipment. Previously, Nokia had produced phones that supported CDMA. At the same time, Qualcomm announced two deals with Chinese companies, giving it an expanded foothold in the potentially huge Chinese market. At the time, that market was dominated by GSM technology.

This chapter addresses the original version of CDMA and its revisions, sometimes known by the trademark cdmaOne. This version of CDMA is governed by specification *TIA/EIA-95-B, Mobile Station-Base Station Compatibility Standard for Wideband Spread Spectrum Cellular Systems (ANSI/TIA/EIA-95-B-99).*

cdma2000 is a 3G evolution of CDMA. To find out more, read the "cdma2000" section in Chapter 10.

Table 8-1 shows the main characteristics of CDMA technology.

TABLE 8-1 **CDMA Technology at a Glance**

Characteristic	Value
Developer	Qualcomm
Initial Deployment	1993
Wireless Generation	2G
Signal Type	Digital
Spectrum Division	CDMA
Frequency	900 MHz, cellular
	1900 MHz, PCS
Channel Spacing	N/A
Switching	Packet-switched
Data Rate	14.4 Kbps

How CDMA Works

While understanding the FDMA technology used in AMPS and the TDMA technology used in GSM is straightforward, understanding CDMA is more difficult. The next few sections go into progressively greater detail on the working of CDMA technology. You won't learn all the high-powered math that is needed to actually implement CDMA technology, but even a passing knowledge will help you spot deployment trends, deploy equipment, and pick service providers that seem to have the best handle on future technologies.

If you want to dig into the math that makes CDMA's black magic possible, visit the CDMA Development Group's Web site, at `www.cdg.org/`. It provides plenty of articles and links to satisfy your curiosity. I recommend you start with an article entitled *The CDMA Revolution,* by Arthur H. M. Ross, Ph.D. (`www.cdg.org/tech/a_ross/CDMARevolution.asp`). The article is a couple of years old, but it provides an introduction to the math behind the technology, as well as an extensive bibliography.

THE 50,000 FOOT VIEW

A good way to attack CDMA is to start with a review of the more straightforward multiple-access technologies: FDMA and TDMA.

FDMA, as its name implies, divides the available radio spectrum by frequency. In other words, there are multiple communication channels, each using a different frequency band, and within any given area, each user gets dedicated use of a frequency band for the duration of a call. Conceptually, this approach is the same as that for a wired phone system, where each call gets a dedicated circuit for the duration of the call. Figure 8-1 provides a graphical representation of the relationship between available frequency bands, time, and users in an FDMA system.

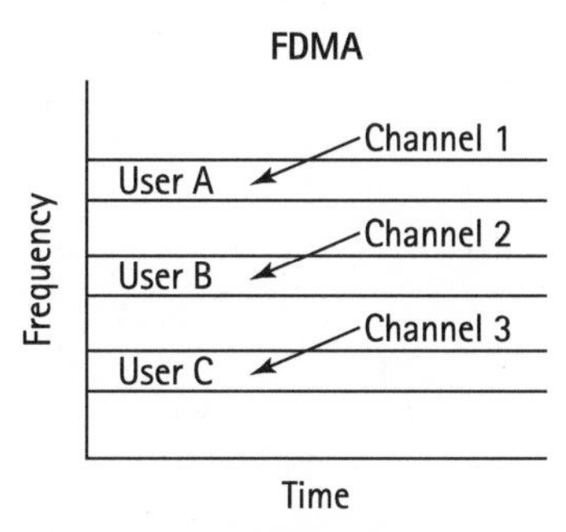

Figure 8-1: The relationship between bandwidth, time, and users in FDMA technology is clear: One user gets one channel for the time it takes to complete the call.

TDMA, on the other hand, divides the available radio spectrum by time. Each TDMA channel can serve multiple simultaneous devices. In a TDMA system, each device on a given channel has a time slot in which it can transmit. Only one device is transmitting on the channel at any given time, but because the time cycles are very short and because each device gets to transmit multiple times per second, this fact is transparent to the user, who perceives uninterrupted use of the channel. Figure 8-2 provides a graphical representation of the relationship between available frequency bands, time, and users in a TDMA system. As you can see from the figure, the relationship is less simple than in FDMA. Here you have several users sharing each channel for the time it takes to complete the call.

CDMA is an entirely different beast. First, CDMA uses digital signals — as does GSM technology in general. But CDMA does something to those digital signals that is, at first glance, bizarre. It uses a *spread spectrum technology* to modulate the digital signal. The actual process involves such intricate equations that Qualcomm obtained a patent on CDMA technology. But to vastly simplify, in effect, each digital signal gets modulated with a high bandwidth signal that looks like so much random noise, smeared across the entire width of the available frequency band, then transmitted on the same channel at the same time as every other CDMA signal in the area! We've tried to represent this in Figure 8-3.

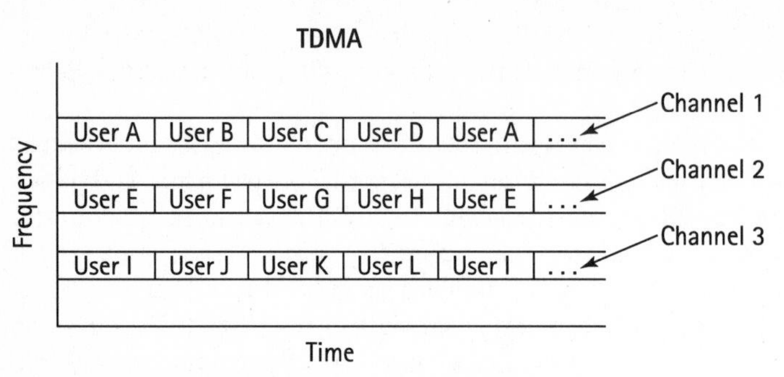

Figure 8-2: The relationship between bandwidth, time, and users in TDMA technology is more complex than in FDMA, but it is still easily comprehensible.

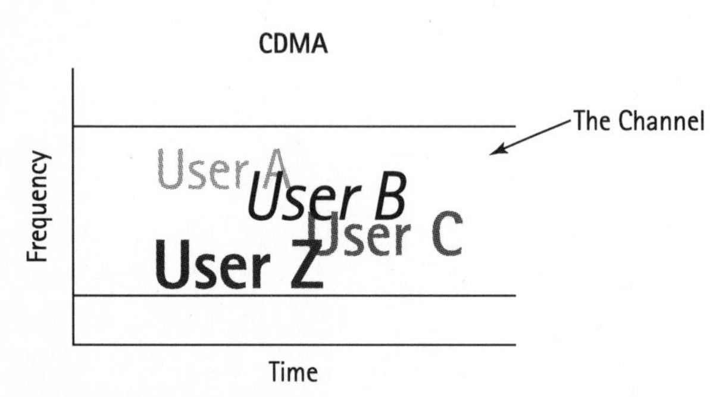

Figure 8-3: The relationship between bandwidth, time, and users in CDMA technology is rather fuzzy and hard to visualize.

Through the use of various black arts (specifically, knowledge of what the carrier signal must have looked like, along with some heavy-duty math, and precision timing signals from Global Positioning System satellites), the receiver extracts the original digital signal. This approach, while complicated and relatively expensive in terms of computing power, yields all the benefits mentioned in the bulleted list at the beginning of this section.

TAKING A CLOSER LOOK AT CDMA TECHNOLOGY

Spread spectrum technology — more specifically, *Direct Sequence Spread Spectrum* (*DSSS)* technology — lies at the core of CDMA technology. As mentioned in the preceding section, DSSS takes the digital signal containing the information you want to transmit and spreads it across a much wider frequency band than would otherwise be required to transmit the signal. This is done by multiplying the digital signal by a digital code that has a much higher frequency than the digital signal, thereby converting the original, relatively low-frequency digital signal into one of high-frequency.

The digital code used to spread the original digital signal is a *pseudo random code,* which appears to be nothing more than a fast, random stream of bits. However, because it is pseudo random instead of truly random, both the transmitter and receiver in a communication system can know exactly what the pseudo random code looks like at any given time.

Once the original digital signal is multiplied by the digital code, the resulting coded signal is used to modulate a carrier wave that is transmitted over the air interface, as in a normal wireless communication system.

When the receiver detects the modulated carrier wave, it demodulates the wave to retrieve the coded signal. Then the receiver multiplies the coded signal by the same pseudo random code used to create the coded signal in the first place. The result of this multiplication is a copy of the original digital signal.

How do the transmitter and receiver manage to use exactly the right pseudo random code at exactly the right time? To achieve precise synchronization between the transmitter and receiver, both must know exactly what time it is. Fortunately, a set of exquisitely precise clocks — the satellites of the Global Positioning System — orbit the earth, broadcasting the exact time and thus giving the transmitters and receivers in a CDMA system the ability to stay synchronized.

So far, sending a digital signal sounds like a lot of work — you might be wondering why anyone would bother to design such a system. But spreading the signal the way DSSS does is the secret to most of the benefits ascribed to CDMA technology. Spread spectrum technology wasn't originally designed for commercial use (see the sidebar "Understanding Spread Spectrum History"). Even so, the characteristics that make spread spectrum so useful to its original designers make it a great commercial technology.

Understanding Spread Spectrum History

While the benefits of the spread spectrum technology used in CDMA are clear, you may be wondering who would develop a modulation technique that involves turning your signal into something that is virtually indistinguishable from noise. The answer is the military.

Spread spectrum technology was developed for the United States military, starting in the 1940s. Back then, the military wasn't interested in cramming lots of mobile phone users into a given radio bandwidth. What attracted the military to spread spectrum was the desire to build a secure, unjammable radio communication system for the armed forces.

This system used spread spectrum radios and — most importantly — satellites to transmit classified information. While the idea of using spread spectrum technology for commercial purposes has been around for decades, it was thought to be impractical because of something known as the *near-far problem.*

The near-far problem for a spread spectrum system is caused by the difference in the signal-to-noise ratio of a signal received from a nearby transmitter and that from a distant transmitter. (For example, when cell phone users are scattered throughout the coverage area of a base station, it may be necessary to have a huge spreading bandwidth to ensure that the more distant transmitters are detected.) The near-far problem was eventually resolved by allowing the base station to control the transmission power of the transmitters it is communicating with such that the received power levels at the base station are roughly equal.

Continued

Understanding Spread Spectrum History *(Continued)*

Spreading a signal across a wide bandwidth gives it the characteristics necessary to create a secure, unjammable system. The best way to protect a signal from prying eyes is to keep the enemy from knowing there is even something to look at. To the uninformed observer, spread spectrum signals appear to be nothing but radio noise. In the case of DSSS, knowing what to look for means knowing the pseudo random carrier wave needed to decode the signal.

As far as jamming goes, the benefits of spread spectrum are clear. The easiest way to jam a normal radio signal is to transmit radio noise on the same frequency as the signal you want to jam. But spread spectrum signals don't work that way. As a matter of fact, the power level of a spread spectrum signal can be below that of the radio noise level — and the message can still be received correctly. Jamming such a signal isn't practical.

Today, the security aspects of DSSS are important to civilians, as well as to the military. The ability of DSSS to function properly in a noisy radio environment will grow ever more valuable as we create millions and then billions of additional wireless devices to share the radio spectrum.

CDMA Benefits

The spread spectrum characteristics of CDMA allow it to make more efficient use of the available radio bandwidth. In general, CDMA can serve eight to ten times the number of users in any given frequency band as AMPS, and around two times the capacity of GSM.

Another CDMA benefit that derives from its spread spectrum nature is its improved noise and interference resistance. When the original digital signals are spread, transmitted, and reconstructed, the total power of noise and interference is dispersed across the entire channel. Thus, bursts of noise or interference (which can pack lots of power in narrow bands) get averaged and become no more than a slight increase in general background noise as far as the spread-spectrum CDMA system is concerned.

System operators directly benefit from the way spread spectrum systems simplify the design of the cellular radio network. As noted in Chapter 7, a typical cellular radio network must divide the available communication channels into seven groups and carefully arrange the physical base stations so that stations using the same channels are not adjacent to each other. CDMA systems use only one channel in any given area and allow all base stations and devices to transmit simultaneously. Since everything uses the same channel, the system operator can deploy base stations wherever needed, regardless of frequency reuse issues.

The section titled "Frequencies and Channels" in Chapter 7 looks at the issue of frequency reuse in some detail.

In Figure 8-4, note that the non-CDMA system (on the left) divides the available bandwidth into seven separate groups of channels, carefully spaced so that adjacent cells do not interfere with each other. In the CDMA system (on the right), all base stations and devices use the same channel.

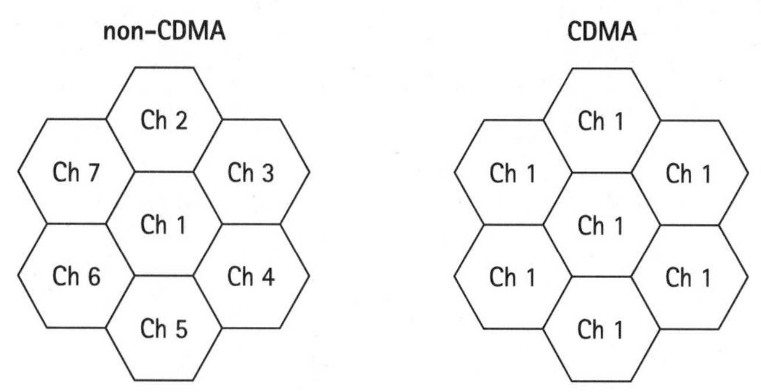

Figure 8-4: CDMA simplifies system planning by allowing all the base stations and devices in an area to share a single channel.

Yet another CDMA benefit that derives from its spread spectrum nature is a reduced tendency to drop calls during handoffs from one cell to another. CDMA systems execute what is called a *soft handoff,* while most other systems execute a *hard handoff.* To hand off a device from one cell to another, most systems must cause the device to switch to another channel, perhaps even another time slot (in TDMA systems). There is some risk that this hard handoff will fail, causing the call to be dropped.

CDMA is inherently more secure than 1G technologies such as AMPS. Normal radio receivers cannot tune the digitally encoded, spread spectrum signals transmitted by CDMA. And since CDMA allows for something over 4 *trillion* unique pseudo random codes, the probability of cloning or other types of fraud is effectively zero.

A problem that affects many wireless systems is *RF multipath,* which occurs when the signals that the wireless device uses take slightly different paths to the receiver. The presence of objects between the transmitter and receiver, and the reflection of the signal from those objects, causes portions of the signal to travel on slightly different paths and thus arrive at slightly different times. In general, RF multipath reduces the performance of a receiver; however, CDMA receivers combine the multiple multipath signals and increase the signal strength.

In a CDMA system, everyone transmits on the same frequency. Transmitters or receivers are only distinguished from one another by the pseudo random code they are using at the moment. CDMA devices can work with two codes simultaneously, so the device can monitor the signal levels and other relevant characteristics of the cell it will change to. Once the device determines that it is time to switch from one cell to another, the device need merely change codes. This results in fewer dropped calls.

The solution to the near-far problem turns out to benefit CDMA users with regard to battery life. To prevent near signals from overwhelming far signals, each wireless device's transmitter should transmit at the minimum power level necessary for communication with the base station at the desired signal-to-noise ratio. Individual transmitter power is controlled by the base station, which in effect reduces the power output from each wireless device as far as it can. As a result, the wireless devices usually transmit at very low power levels, giving them longer talk times and standby times than other wireless technologies.

Digging Even Deeper into CDMA and Spread Spectrum

If you want to dig even deeper into CDMA and the spread spectrum technology it is built on, you'll have to journey to different sources. Once again, I suggest you start at the CDMA Development Group Web site (www.cdg.org/), with *The CDMA Revolution*, by Arthur H. M. Ross, Ph.D. (www.cdg.org/tech/a_ross/CDMARevolution.asp). Another good resource is the Qualcomm CDMA Development Group Web site at www.qualcomm.com/cda/technology/display/0,1704,,00.html. Here, you can not only read about CDMA and spread spectrum technology but also arrange for special classes with Qualcomm's CDMA University.

For the latest and greatest information on CDMA, visit the Web site of the CDMA Development Group (www.cdg.org). This multilingual site provides many articles, links, and other resources related to CDMA. The organization is "an international consortium of companies who have joined together to lead the adoption and evolution of CDMA wireless systems around the world." So check it out if you want the latest information on CDMA.

Cellular Digital Packet Data

CDPD stands for *Cellular Digital Packet Data*. CDPD is a digital *packet* data technology that can be deployed on top of existing AMPS and TDMA networks at relatively little cost. CDPD offers numerous benefits for applications such as wireless e-mail and World Wide Web access, and most United States AMPS providers have made CDPD available in the majority of their networks. This technology is also known as *Wireless IP (Wireless Internet Protocol)*.

You can find online maps of CDPD coverage areas. Your best bet is to check the Web sites of the mobile phone companies covering your area. You can also visit the CTIA's wireless coverage maps page at www.wirelessdata.org/maps/index.asp. Some of these maps are a little dated, but with links to a many different maps and coverage-related resources all in one place, it is worth your while to visit this site if you're thinking about CDPD coverage.

CDPD technology was developed by the member companies of the *Wireless Data Forum (WDF)*. The WDF has since merged with the *Cellular Telecommunications and Internet Association (CTIA)*.

Many of the most popular wireless add-ons for handheld devices rely on CDPD to transfer their data. Both the OmniSky/Earthlink and YadaYada (currently in financial straits) wireless service providers (WSPs) build their solutions around CDPD modems from Novatel Wireless. Likewise, wireless modems from Enfora and Sierra Wireless run on the CDPD network.

In December of 2001, both OmniSky and YadaYada filed for bankruptcy protection under Chapter 11.

You can find more information on wireless modems from Novatel Wireless, Enfora, and Sierra Wireless in Chapters 2 and 3.

Table 8-2 shows the main characteristics of CDPD technology.

TABLE 8-2 **CDPD Technology at a Glance**

Characteristic	Value
Developer	Member companies of the WDF (now merged with the CTIA)
Initial Deployment	1993
Wireless Generation	2G
Works with	AMPS and TDMA cellular networks
Security	Encryption and channel hopping
Supported Standards	IP, CLNP, IP Multicast, IPv6
Error Correction	Forward Error Correction
Switching	Packet-switched
Data Rate	19.2 Kbps

How CDPD Works

The data that CDPD carries is digital, yet it travels on the analog AMPS network. Therefore, you must install one or more digital radios in order to add CDPD to an AMPS base station. The combined system must somehow support both analog and digital data. Add to this the fact that — according to its design requirements — CDPD must be compatible with existing systems and have minimal impact on end systems, and you are left with an interesting design problem.

CDPD works with packets of data. A packet of data consists of a quantity of digital data, along with routing information and some error detection and correction data. It turns out that the exact format and addressing scheme used in these packets is determined by the Internet protocol (hence the use of the term "Wireless IP" to refer to CDPD).

Building CDPD on Existing Infrastructure

An important design goal for CDPD was to be able to build on the large, existing AMPS infrastructure. As described at the CTIA's Web site, "The CDPD Network is designed to operate as an extension of existing data communications networks. This extension is provided through definition of the CDPD Network as a peer multi-protocol, connectionless network to existing data infrastructures. Existing connectionless network protocols may be used to access services through the CDPD Network. Similarly, existing connectionless network protocols may be used to access services provided by the CDPD Network."

To meet this goal, the system design specification for CDPD imposed requirements such as the following:

- ✓ To ensure compatibility with existing systems
- ✓ To allow a phased deployment strategy
- ✓ To minimize impact on end systems

The end result meets these requirements, allowing the service providers to add CDPD capability to their base stations when and where they see fit, with no discernible impact on their existing AMPS service.

If it interests you, the CDPD System Specification Release 1.1 consists of eight Books covering five Fasciles (or editions), each of which includes a number of Parts. There is also a CDPD Implementer's Guide, which consists of two Books, collectively containing five Fasciles.

To summarize, CDPD must take digital data, break it up into packets, and move those packets across the AMPS network without any noticeable impact on that network. To meet these requirements, the designers of CDPD added a frequency-hopping capability to their digital radios.

In a nutshell, the frequency hopping works as follows: When a device or base station is ready to transmit a packet of CDPD data, it scans the AMPS channels, looking for one that is idle. In other words, it looks for a channel that is not in use at that moment, or even one that is normally reserved for use during handoffs.

Once the CDPD digital radio finds an idle channel, it begins transmitting data packets. At the same time, it monitors the channel it is presently transmitting on, to see if someone has activated an AMPS transmitter on the same channel. If so, the transmitter hops to a new unused channel and resumes sending packets. Meanwhile, the receiver in the CDPD digital radio is scanning all the AMPS channels, looking for any packets that are addressed to it. When it finds such packets, the receiver uses the rules codified in the Internet protocol to check and extract the data in the packets and reassemble the original message.

Packet-Switching, Frequency-Hopping Benefits

Making CDPD a frequency-hopping, packet-switched technology offers many benefits besides the ability to piggyback a load of digital data on top of AMPS. From the user's perspective, CDPD appears to be continuously connected. That is, you don't need to dial an access phone number or wait for a circuit to be set up as you do when making a voice call or using an analog modem. When the device is ready to send a packet, it finds an open channel and sends the packet. Likewise, the device can continuously scan for incoming packets that are addressed to it.

Thus, using a CDPD-equipped device to transfer data feels more like using a (very slow) computer hooked up to a network, rather than calling someone on the phone.

Similarly, a CDPD-equipped device doesn't tie up an entire channel for the entire time the device is active. Each device only uses the channel when it is actually sending a packet. The rest of the time, the channel is available for use by other CDPD devices, or for regular AMPS use.

Since CDPD creates packets based on the Internet protocol, service providers can easily give CDPD users access to the Internet. While no one is going to be surfing the Web with a CDPD phone, easy Internet access offers tremendous possibilities for delivering priority data to the user when and where it is needed.

More CDPD Benefits

In addition to the benefits derived from its packet-switched, frequency-hopping nature, CDPD offers the following benefits:

✓ Using Reed-Solomon forward error-correcting codes ensures data integrity. With Reed-Solomon codes, a CDPD system can detect and correct errors in blocks of digital data. Forward error-correcting is obviously of great value in the wireless environment, where the air interface is subject to a large amount of interference, and a low data transfer rate means that retransmitting data should be avoided.

For more detailed information on Reed-Solomon codes, check out this 1993 article originally published in the SIAM News (SIAM stands for Society for Industrial and Applied Mathematics): www.siam.org/siamnews/mtc/mtc193.htm.

✓ Since CDPD works on the existing AMPS network, a dual-mode CDPD/AMPS phone can share most of the electronics needed to function in both modes.

✓ The CDPD System Specification calls for authentication and encryption of all packet transmissions, thereby protecting the privacy of the user and greatly reducing the chance of someone impersonating or cloning a user's account.

Global System for Mobile Communication

GSM, or *Global System for Mobile Communication*, is by far the most popular system for cellular communications worldwide. By the spring of 2001, the total number of GSM subscribers worldwide had already exceeded half a billion. GSM is the de facto wireless standard in Europe, and is widely used throughout the world (Japan being the notable exception). In Japan, the DoCoMo i-mode service (built on the PDC-P standard) dominates.

The GSM Association is the international body behind GSM. It is responsible for the "development, deployment and evolution of the GSM standard for digital wireless communications and for the promotion of the GSM platform." In June 2001, the GSM Association announced "the creation of the Mobile Services Initiative (M-Services), an unprecedented global industry move to enhance benefits to consumers using GSM handsets by delivering a globally available set of services through the mobile Internet."

M-Services is a set of guidelines for services that can be offered by any GPRS handsets. Services mentioned in the initial press release include enhanced graphics, music, video, games, ring tones, and screen savers. The impact of this announcement is hard to predict, but you can get the latest news on M-Services and GSM in general at the GSM Association's Web site, GSM World: `www.gsmworld.com`.

In the United States, providers using GSM serve a growing proportion of the overall wireless community. AT&T Wireless is in the process of replacing its TDMA (D-AMPS) services with GSM, and Cingular Wireless is expected to follow suit. These moves will greatly increase the availability of GSM service in the U.S., and will likely sound the death knell for TDMA as a major telecommunications service.

If you want to see where GSM coverage is available in any country on earth, you are in luck. `GSMCoverage.co.uk` provides links to GSM coverage maps for the entire planet. Webmaster Mark Paulton maintains this extensive site as a hobby, providing a valuable service to the Internet community. The site can be found at `www.gsmcoverage.co.uk/`.

Given the sheer number of GSM users, and GSM's impact on the telecommunications market, this chapter provides more information about GSM than about any other 2G technology.

Table 8-3 shows the main characteristics of GSM.

TABLE 8-3 **GSM at a Glance**

Characteristic	Value
Responsible Party	European Telecommunications Standards Institute (ETSI)
Initial Deployment	1991
Wireless Generation	2G
Signal Type	Digital
Spectrum Division	FDMA/TDMA
Frequency	GSM-900, 890–960 MHz
	GSM-1800, 1710–1880 MHz
	GSM-1900, 1850–1990 MHz
Channel Spacing	200 KHz
Data Rate	9.6 Kbps (per channel)

How GSM Works

GSM is a digital technology that works by dividing the available radio spectrum into discrete channels (the FDMA approach). GSM further divides each channel into eight time slots (the TDMA approach), allowing up to eight users to occupy a channel at a time. The GSM recommendations define a system that carries voice and data, includes sophisticated error detection and correction, and provides high security and privacy.

GSM systems allow for worldwide roaming, making it easy for users to roam from system to system using their wireless devices wherever they are.

A typical GSM network is composed of several subsystems, including the *Mobile Station*, *Base Station Subsystem* (*BSS*), *Network and Switching Subsystem* (*NSS*), and *Operating and Support Subsystem* (*OSS*). Figure 8-5 illustrates these subsystems, as well as some of their components.

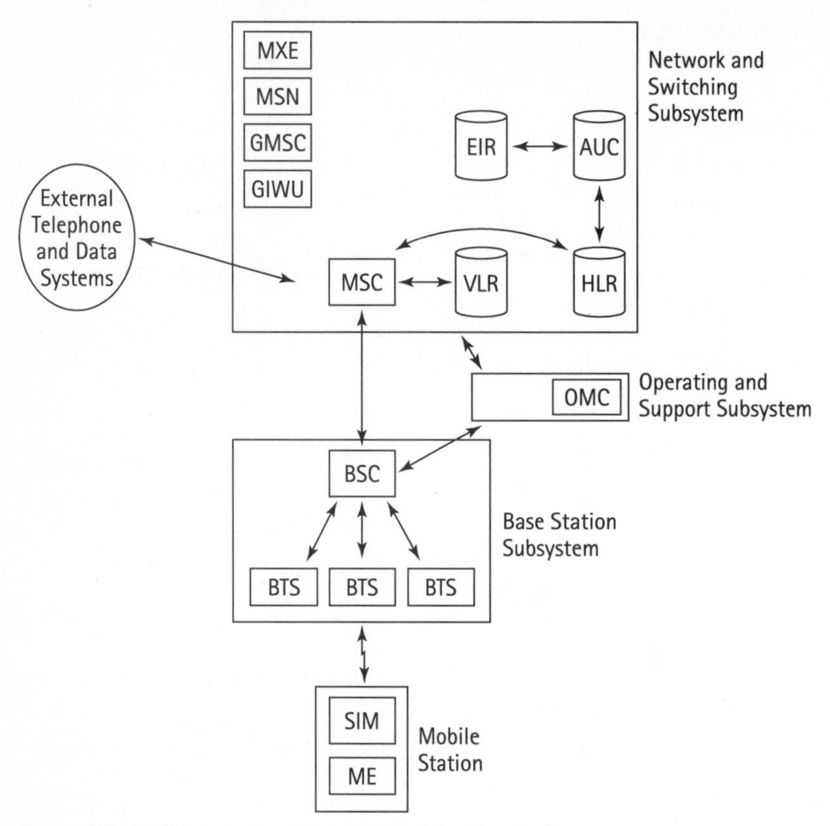

Figure 8-5: A GSM network consists of several functional subsystems.

MOBILE STATION COMPONENTS

The mobile station consists of two components:

- ✓ **Subscriber identity module (SIM):** The SIM is a smart card that contains the information needed to identify a subscriber, connect that subscriber to the network, and ensure that the subscriber can be billed for service. The contents of the SIM include the subscriber's unique *International Mobile Subscriber Identity* (*IMSI*). The SIM provides personal mobility for GSM subscribers. By inserting the SIM into the appropriate mobile equipment, the user can connect to a GSM network, regardless of the specifics of that network.
- ✓ **Mobile equipment (ME):** The ME is the physical hardware that provides the connection to a GSM network (most commonly, a mobile phone). Each unit of mobile equipment is uniquely identified by an *International Mobile Equipment Identity* (*IMEI*).

BASE STATION SUBSYSTEM COMPONENTS

The BSS consists of two types of components:

- ✓ **Base station controller (BSC):** Each BSC controls one or more BTSs. Specifically, the BSC is responsible for the channel selection, frequency hopping, and *handovers* (handoffs) of the mobile stations communicating with any of the BTSs it controls.
- ✓ **Base transceiver station (BTS):** Each BTS contains the radio transceivers and antennas needed to service one cell of the GSM network. One or more BTSs is controlled by each BSC.

NETWORK AND SWITCHING SUBSYSTEM COMPONENTS

The NSS consists of a variety of components:

- ✓ **Authentication center (AUC):** The AUC is a protected database that provides authentication and encryption information used to confirm a user's identity and ensure the confidentiality of a call. The AUC can provide this information because it holds a copy of the secret key stored in each subscriber's SIM.
- ✓ **Equipment identity register (EIR):** The EIR is a database that contains the IMEI for every ME on the network. If any ME is reported stolen, or its type is no longer authorized on the network, it is marked invalid in the EIR and prevented from using the network.
- ✓ **GSM interworking unit (GIWU):** The GIWU is an interface between various data networks. It provides certain capabilities, such as the ability to switch between voice and data transmission while a call is in progress.
- ✓ **Gateway mobile services switching center (GMSC):** An MSC can serve as a gateway between the GSM network and other networks. When it does so, it is referred to as a GMSC.
- ✓ **Home location register (HLR):** The HLR holds information about subscriptions to the GSM network. This information includes each subscriber's service profile, location information, and activity status. The location information is in the form of a signaling address, not latitude and longitude, or any other type of measurement that would indicate a person's exact location. Conceptually, there is only one HLR for the entire network, although in practice, the database may be distributed to multiple locations as required.
- ✓ **Mobile services switching center (MSC):** The MSC is the heart of the NSS. It performs the telephony switching functions for the system, routing calls as necessary. It also provides the mobile user management functions of the system. In particular, the MSC (with the help of several other NSS components) manages subscriber registration and authentication, tracks the subscriber's location, and supervises handovers.
- ✓ **Mobile services node (MSN):** The MSN handles mobile intelligent network functionality for the MSC. One example of the kind of functionality that could be provided through the MSN is a mobile virtual private network (mobile VPN).
- ✓ **Message center (MXE):** The MXE provides voice, data, and fax messaging services to the MSC. These include SMS, cell broadcast, e-mail, voice mail, and fax services.

- ✓ Visitor location register (VLR): The VLR contains a subset of the information in the HLR. Furthermore, it only holds such information for users currently within the geographical area controlled by the VLR. Information from the VLR is provided to the MSC to enable it to do its job. In most cases, the system is designed so that the area controlled by the MSC and the VLR is the same.

For the purposes of this discussion, assume that the GSM network under study contains a single *mobile services switching center (MSC).*

OPERATION AND SUPPORT SUBSYSTEM COMPONENTS

The OSS contains only a single component, along with that component's connections to the rest of the GSM system. The single component of the OSS is the *operations and maintenance center (OMC).* The OMC provides for centralized management of the major components of a GSM network. It has connections to all the components of the NSS and to the BSC. An OMC provides a network with fault detection, configuration management, performance management, and security management features.

Understanding GSM Network Areas

GSM networks, like all other 2G wireless services, provide coverage across a certain geographic area. This geographic area is subdivided into four levels, each of which is reflected in the logical structure of the GSM network. What follows is a description of each of the four levels, starting at the smallest — the individual cell.

CELLS

The lowest level of the GSM network is the individual cell. Each cell is served by a single BTS, and the boundaries of the cell are determined by the size of the area the BTS covers. Every cell in the network has a *cell global identity number* assigned to it.

LOCATION AREAS

A *location area* (*LA*) consists of a group of cells, controlled by at least one BSC, but only one MSC. Figure 8-6 illustrates this relationship. When the system pages a user, the page is sent to the location area that the system believes the user to be within.

MSC/VLR SERVICE AREAS

Moving up one level, the *MSC/VLR service areas* are areas that are controlled by a single MSC or VLR. The use of multiple location areas is an example of why it makes sense to have MSCs and VLRs cover the same areas, as the logic for these kinds of areas would become needlessly complex if they did not. As Figure 8-7 shows, an MSC/VLR service area can contain multiple location areas, and therefore many cells.

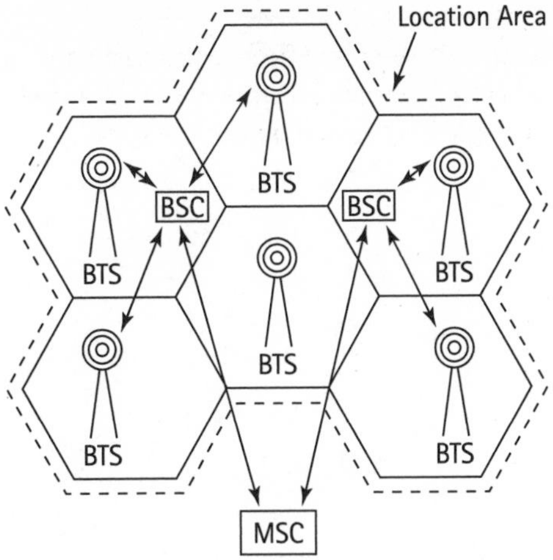

Figure 8-6: A GSM location area consists of several cells, all controlled by a single mobile services switching center (MSC).

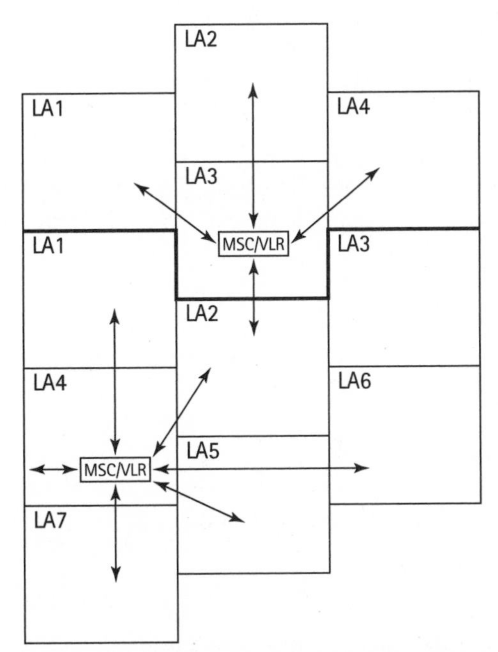

Figure 8-7: A GSM MSC/VLR service area consists of one or more location areas, all controlled by a single MSC (and its associated VLR). This figure shows two service areas.

PLMN SERVICE AREA

The *public land mobile network (PLMN) service area* is the largest area within a GSM network. It is the total area served by that network (see Figure 8-8). A PLMN service area can contain as many MSC/VLR service areas as needed to cover the total geographic area served by the network.

Figure 8-8: A PLMN service area consists of one or more service areas, controlled by as many MSCs (and their associated VLRs) as it takes.

GSM Implementation Details

Now that you understand the basic architecture of a GSM network, and the major components that comprise that network, it is time to look at some of the implementation details. These include the basic workings of the air interface, error detection and correction, and handling handovers.

GSM AIR INTERFACE

The GSM air interface operates in three different frequency bands: 900 MHz (the original GSM), 1800 MHz (formerly known as DCS 1800), and 1900 MHz (GSM in the United States).

While operating on three different frequency bands, each system is still functionally the same. The available bandwidth is divided into channels using FDMA. The channels are divided into two groups, with a channel from each group combined into a pair. A call of any sort involves both channels in the pair.

The higher frequency channel in the pair carries transmissions from the BTS to the ME. The lower frequency channel in the pair carries transmissions from the ME to the BTS. In this respect, GSM is similar to the 1G TACS system.

TACS stands for Total Access Communication System. It is a 1G cellular system deployed across the United Kingdom and several other countries. See "TACS/ETACS" in Chapter 7 for more information.

GSM 900 supports 124 paired channels, GSM 1800 supports 374 paired channels, and GSM 1900 supports 299 paired channels. GSM transmits digital signals on its channels, instead of analog signals like TACS. Furthermore, each of the channels is divided into time slots using TDMA technology. The GSM recommendations show eight full rate time slots in each channel, effectively allowing eight users to share a channel simultaneously.

While the preceding explanation of how a channel is divided up using TDMA is effectively correct, it leaves out many details. If you look closely at the channels, you find that the signals on a channel are actually divided into 26-frame multiframes. Each frame is divided into eight burst periods of 0.577-milliseconds duration. Each burst period is further subdivided into odd units like two sets of Tail Bits, two sets of 57 Data Bits, a Stealing Bit, a Training Sequence, and 8.25 Guard Bits. Unless you are designing GSM hardware, you should stick with the idea that GSM uses TDMA technology to allow each channel to support eight users.

Given that each channel can support eight simultaneous users, the total number of users that can be supported by GSM 900 is 992, by GSM 1800, 2992, and by GSM 1900, 2392.

CELLS AND FREQUENCY REUSE The preceding calculations for the maximum number of users for each type of GSM do not take into account that GSM, like all the other 1G and 2G technologies covered so far, supports cells and frequency reuse. Thanks to GSM's strong error-correction capabilities, a GSM system can work with a frequency reuse factor of 4, as opposed to the reuse factor of 7 that is common for AMPS and TAC. The strong error correction is necessary because, all other things being equal, the distance between cells using the same frequencies is less with a reuse factor of 4 than with a reuse factor of 7. Figure 8-9 shows cell patterns with reuse factors of 4 and 7.

The reduced reuse factor of GSM enables each of the cells (specifically the base transceiver station for that cell) to use more of the available channels, thereby allowing each cell to support far more simultaneous users than would be possible with a reuse factor of 7.

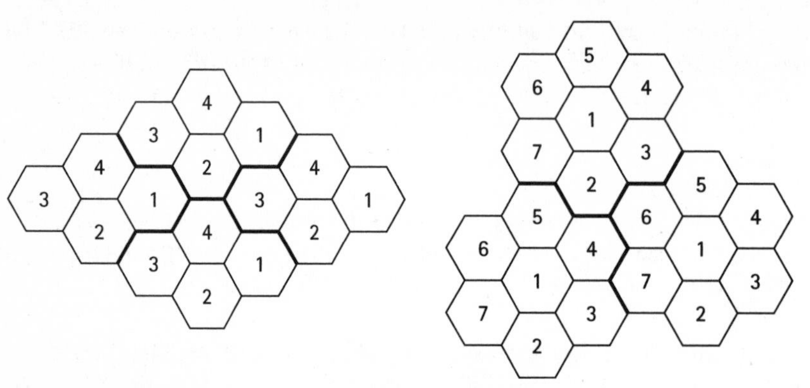

Figure 8-9: A frequency reuse factor of 4 (left) and a frequency reuse factor of 7 (right)

ADDITIONAL ASPECTS OF THE AIR INTERFACE As if the air interface details just discussed weren't complex enough, a number of additional factors affect how a digital GSM signal gets transmitted and received.

- ✓ Without precautions, the signals from one burst period can interfere with those of the following burst period, during which another set of mobile equipment is transmitting. To prevent this problem, the transmitter power must be ramped down at the end of a burst period and ramped up again at the beginning of the next burst period. GSM provides for fifteen steps of power control, with each step corresponding to a change in transmitted power of 2dB.
- ✓ Multipath equalization is needed in GSM systems, particularly GSM 900. The ME determines the effect of *multipath fading* on a known signal (the Training Sequence mentioned earlier) then computing a filter to effectively remove the fading from the user data signal.
- ✓ *Frequency hopping* can be used in a GSM system to minimize interference, alleviate multipath fading, and reduce the interference between channels. The GSM recommendations provide for 217 frequency hops per second. The algorithm used to determine which frequency will hop to where next is broadcast on one of the control channels, specifically the *broadcast control channel.*
- ✓ GSM systems must be able to maintain communications with mobile equipment in vehicles moving at up to 250 kilometers per hour, compensate for the propagation delay when signals travel long distances, and handle frequency standard drift.
- ✓ To save power, GSM mobile equipment can implement *discontinuous transmission.* Research has shown that the typical person talks only 40 percent of the time during an average phone conversation. Discontinuous transmission takes advantage of this fact by turning off the transmitter during the time that the user is not actively talking.

- ✓ *Discontinuous reception* is another technique that can save power. It is only possible because the mobile equipment knows when its next time slot should arrive. When the mobile equipment receiver isn't actively listening for data, it turns itself off, thus saving power.

HANDOVERS

Handovers are one of the most important aspects of call handling. GSM systems recognize four types of handover, each of which has its own characteristics and requirements. The four types of handover in a GSM system are the following:

- ✓ Handovers between channels and time slots controlled by the same base transceiver station.
- ✓ Handovers between base transceiver stations controlled by the same base station controller.
- ✓ Handovers between base transceiver stations controlled by different base station controllers, but the same mobile services switching center.
- ✓ Handovers between base station controllers controlled by different mobile services switching centers.

The first two types of handovers are known as *internal handovers*. Internal handovers can be managed completely by the BSC, which needs only to notify the MSC that the handover has occurred. This eliminates unnecessary communication between the BSC and the MSC.

The last two types of handovers, those involving different BSCs controlled by the same MSC and those involving different MSCs, are *external handovers*. External handovers must be managed, at least partially, by an MSC.

In GSM systems, the MSC that manages the BSC that manages the BTS that originally contacted the mobile equipment (since the ME last connected to the GSM system) is known as the *anchor MSC*. The anchor MSC maintains the subscriber's information and call-related functionality. When a call is handed over to another MSC, that MSC is known as the *serving MSC*, or *relay MSC*. The relay MSC manages subsequent inter-BSC handovers, but responsibility for most of the subscriber's information and functionality remains with the anchor MSC.

Either the mobile equipment or the controlling MSC can initiate a handover. The MSC initiates handovers as a means of balancing the traffic load on the system. The mobile equipment initiates a handover based on the algorithms programmed into it. These algorithms are not defined by the GSM recommendations, but are designed by the mobile equipment manufacturers.

When it is time to initiate a handover, the mobile equipment scans the broadcast control channel to find the channels of up to 16 neighboring cells. From this list, the mobile equipment determines the best six handover candidates, based on received signal strength from each. The mobile equipment then passes this information on to the BSC or MSC, which uses it to decide which channel the mobile equipment should hand over to.

Determining When to Hand Over a Call

The mobile equipment can initiate a handover when the power level of the signals it receives declines below a threshold value. The signal power can be reduced when the mobile equipment moves to the edges of a cell, or when the signal is affected by multipath interference. While the GSM recommendation doesn't specify the algorithms that mobile equipment should use to initiate a handover, there are at least two classes of algorithm in use:

- ✓ **Minimum acceptable performance:** The minimum acceptable performance type of algorithm attempts to minimize handovers by increasing the mobile equipment's power level when the incoming signals degrade to a certain point. This method is easier to implement and, apparently, more popular than the power budget type of algorithm. It does, however, have some drawbacks: potentially greater power use (hence shorter battery life) and the possibility of causing interference due to increased transmission power levels.
- ✓ **Power budget:** The power budget type of algorithm attempts to maintain a specified signal quality while keeping the power level as low as possible. This approach leads to more handovers and is more complicated to implement. The advantages of the power budget type of algorithm include potentially greater battery life and a reduced possibility of causing interference.

Services Provided by GSM Systems

GSM systems support a wide range of services for subscribers. Some, called *subscriber services*, are provided as part of the basic service subscribers get with their GSM subscription. Others, called *supplementary services*, are services that subscribers can use to supplement their basic service for an additional fee.

The voice-related services, whether subscriber services or supplementary services, are sometimes called *teleservices.* The data-related services, whether subscriber or supplementary, are sometimes called *bearer services.*

SUBSCRIBER SERVICES

The following list details the components of basic service in a standard GSM subscription.

- ✓ Basic voice telephone: This includes standard voice telephone service and may or may not include features such as call waiting or three-way calling.
- ✓ Emergency calling: A subscriber should be able to initiate an emergency call, taking bandwidth priority away from other non-emergency calls.

- **Cell broadcast:** This service enables the user to send a short (93-character) text message to everyone within a particular cell. It operates in the background, so the mobile equipment only accepts cell broadcast messages when it is idle. Cell broadcast is often used for traffic warnings or to report accidents. The subscriber can specify which types of cell broadcast messages (weather, but not traffic reports, for example) the mobile equipment will accept.
- **Dual tone multi-frequency (DTMF) tones:** DTMF service enables GSM subscribers to control devices, such as home answering machines, the same way they would using a touch-tone phone — by pressing the keys on the handset in the proper pattern.
- **Facsimile group III:** This service enables subscribers to communicate with standard analog fax machines to send and receive faxes through the GSM network.
- **Fax mail:** This service allows subscribers to store faxes in a GSM service center and retrieve them at any fax machine. To protect the privacy of these stored faxes, the subscriber must enter a personal security code before retrieving stored faxes.
- **Short message service (SMS):** SMS is one of the most popular features of GSM phones, and has become one of the most popular communication technologies in existence, with literally billions of SMS messages being sent every month. See Chapter 11 for more information.
- **Voice mail:** Voice messages for the subscriber can be stored in a voice mailbox somewhere in the GSM network. The subscriber's messages are protected by a personal security code.

SUPPLEMENTARY SERVICE

The GSM recommendations define a number of supplementary services that should be supported by a GSM network. This number is subject to change as new services are developed, but the supplementary services that are usually offered include the following:

- **Advice of charge (AOC):** This service gives the subscriber an estimate of the charges for a particular call. It works with both teleservices and bearer services.
- **Barring of incoming calls:** This service prevents the mobile equipment from receiving calls. It might be used when the subscriber is traveling outside the home GSM network, thereby helping him or her avoid roaming charges.
- **Barring of outgoing calls:** This service allows the subscriber to prevent the mobile equipment from making calls, helping prevent unauthorized phone use.
- **Call forwarding:** This service allows subscribers to specify a number for calls to be forwarded to, and the conditions under which calls should be forwarded (for example, when the mobile equipment is out-of-service or if there is no reply to a call).
- **Call hold:** This service allows subscribers to put a call "on hold." It only applies to voice calls.
- **Call waiting:** This service notifies the subscriber of an incoming call while another call is in progress. Call waiting is applicable to voice calls.

- **Calling line identification:** This service displays the ISDN number of a caller on the mobile equipment's display without accepting the call.
- **Closed user groups:** This service makes it possible to define groups of subscribers who can only call certain phone numbers.
- **Multiparty service:** This service allows subscribers to set up a multiperson call, with as many as three to six subscribers dialing in. Only voice calls can be multiparty.

As the preceding pages have shown, GSM is a complex, feature-rich, and highly popular technology. While it is "only" a 2G technology now, significant research efforts have been made to move the essence of GSM onto 2.5G and 3G technologies as they are deployed. Expect GSM and its descendants to play an important role in the foreseeable future.

Personal Communications Services

PCS stands for *Personal Communications Services*. PCS is not really a technology. Instead, it is a creation of the United States Federal Communications Commission (FCC), broadly defined as "radio communications that encompass mobile and ancillary fixed communication services that provide services to individuals and businesses and can be integrated with a variety of competing networks." To facilitate the development of PCS, the FCC selected the radio spectrum between 1850 MHz and 1990 MHz as the location for PCS, divided it into six licensed bands and one unlicensed band, and auctioned the licensed bands in various areas of the country to the highest bidder.

To further muddy the waters, the version of PCS that occupies 1850–1990 MHz is actually classified by the FCC as "broadband PCS." There is also something called "narrowband PCS" — three 1-MHz bands set aside for PCS in the 900 MHz range (901–902 MHz, 930–931 MHz, and 940–941 MHz). However, when people talk about PCS, they are almost always referring to broadband PCS, so there will be no further mention of narrowband PCS in this chapter.

Table 8-4 shows the main characteristics of PCS.

TABLE 8-4 **PCS at a Glance**

Characteristic	Value
Developer	Federal Communications Commission (USA)
Initial Deployment	1995
Wireless Generation	2G

Characteristic	Value
Signal Type	Digital
Spectrum Division	TDMA, CDMA
Frequency	1850–1990 MHz
Channel Spacing	Varies
Switching	Varies
Data Rate	Varies

Once you understand that PCS is a concept and a set of frequency bands, instead of a specific technology, much of the confusion goes away. Specific PCS technologies operate the same way they do in other frequency bands. For example, note the following:

- ✓ They use cells.
- ✓ They perform handoffs when users move between cells.
- ✓ They employ frequency reuse if they are built on CDPD or GSM technology.
- ✓ They do not employ frequency reuse if they are built on CDMA technology.

Expert Advice: Divining the Difference between PCS and Cellular

If you were to pay attention to radio and television commercials promoting PCS, you would be forgiven for thinking there were clear and unarguable differences between PCS and cellular systems. As far as technology goes, there aren't any. In the United States, GSM service is provided in the PCS band. There are some CDPD services in the PCS band, and Sprint PCS, which operates the largest PCS network in the United States, uses CDMA technology.

On the other hand, some people consider any digital phone system that supports data and voice to be PCS, no matter what frequency band that system uses.

We suggest that when you talk about PCS, you ignore the specific technology and focus instead on the kinds of services it provides and the frequency band being used. The FCC set up PCS to be technology transparent, which helps avoid confusion.

The PCS frequency band is the only significant topic that needs to be discussed. Table 8-5 identifies the six licensed bands and one unlicensed band reserved for PCS use. Three bands are 30 MHz wide (15 MHz from device to base station, 15 MHz from base station to device), three are 10 MHz wide (5 MHz from device to base station, 5 MHz from base station to device), and the unlicensed band is 20 MHz wide (just one big chunk of spectrum, completely undivided).

TABLE 8-5 **PCS Frequency Band Breakdown**

Band	Licensed?	Lower Portion of Band	Upper Portion of Band
A	Licensed	1850 MHz–1865 MHz	1930 MHz–1945 MHz
B	Licensed	1870 MHz–1885 MHz	1950 MHz–1965 MHz
C	Licensed	1895 MHz–1910 MHz	1975 MHz–1990 MHz
D	Licensed	1865 MHz–1870 MHz	1945 MHz–1950 MHz
E	Licensed	1885 MHz–1890 MHz	1965 MHz–1970 MHz
F	Licensed	1890 MHz–1895 MHz	1970 MHz–1975 MHz
Unnamed	Unlicensed	1910 MHz–1930 MHz	1910 MHz–1930 MHz

Interestingly enough, nothing in the definition of PCS prevents faster, more capable technologies (2.5G or 3G) from being deployed in the same frequency bands, and under the PCS description. Thus, don't be confused if some day soon you start hearing about high-speed PCS, or PCS2, or another buzzword associated with a faster and more capable version of PCS. It will likely just be some new digital wireless technology, deployed in the 1900 MHz band and marketed under a name people already recognize.

Time Division Multiple Access

TDMA (formerly known as D-AMPS) is a 2G digital upgrade to the 1G AMPS technology that is widely used in the United States. Devices that use TDMA can usually also work with the older AMPS technology in areas where there is AMPS, but not TDMA, coverage. Many mobile phones advertised as "dual band dual mode" are actually D-AMPS phones that can also communicate on regular AMPS when necessary. TIA/EIA-136 is the Telecommunication Industry Association standard that defines TDMA.

Table 8-6 shows the main characteristics of the TDMA wireless communication technology.

TABLE 8-6 **TDMA Technology at a Glance**

Characteristic	Value
Developer	Ericsson
Initial Deployment	1991
Wireless Generation	2G
Signal Type	Digital
Spectrum Division	TDMA
Frequency	824–894 MHz
Channel Spacing	30 KHz
Switching	Circuit-switched
Data Rate	9.6 Kbps

Just because a mobile phone is dual mode dual band, and can connect in areas that do not have TDMA coverage, doesn't mean you will get the same kind of service operating on AMPS as you would on TDMA. When in AMPS, most or all of the cool digital features on a phone (e-mail and Web browsing, for example) typically won't work. The phone will be able to make analog (1G) voice calls, and little else.

TDMA works by using a time-division multiplexing scheme to transmit three digital channels in the same frequency band as one old-fashioned AMPS channel. With this approach, much of the existing AMPS infrastructure can be reused. The phone companies can gradually add TDMA capabilities to base stations and transition from AMPS to TDMA without the massive up-front costs of deploying an entirely new cellular radio network all at once.

Some clarification is needed regarding the term *TDMA.* TDMA, which stands for *Time Division Multiple Access,* is a technology that divides available channels by time, allowing each transmitter to transmit only at certain times. Thus, you end up with a wireless communication technology called *TDMA,* which divides up its available spectrum using a technology called *TDMA.* Even worse, a competing wireless communication technology, GSM, uses TDMA to divide its available spectrum. In the rest of this section, we will indicate which TDMA we are talking about — if not clear from the context.

TDMA supports a technique known as *hierarchical cell structure*, or *HCS*. HCS recognizes three types of cells: *macro*, *micro*, and *pico*. Micro cells can exist within macro cells. Likewise, pico cells can exist within macro and micro cells. Don't confuse this hierarchical division with cell splitting. In cell splitting, an entire cell is divided into a set of smaller cells, as shown in Figure 8-10.

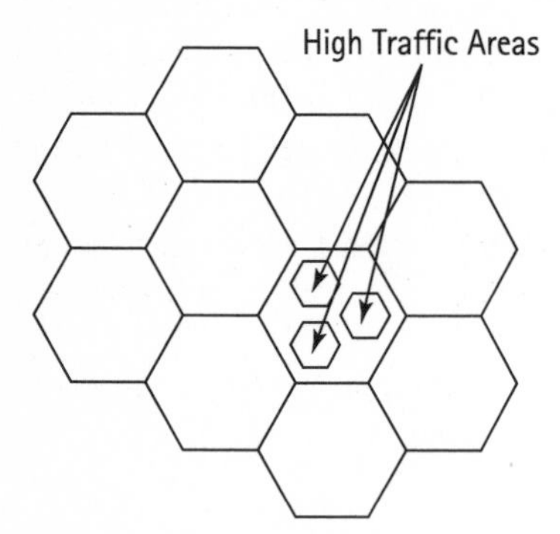

Figure 8-10: Cell splitting divides a cell into a set of smaller cells.

HCS is a more flexible approach than cell splitting. With this technique, smaller cells can exist within larger cells. They can be placed in sections of the larger cell that will likely experience high cellular traffic. This technique could profitably be used in a high-tech business park in an undeveloped area. Such a location would likely contain a high concentration of cellular users (the park itself), surrounded by a much larger area likely to contain a low concentration of users (the parking lots and undeveloped lands around the park). With an HCS system, engineers could create many micro (or even pico) cells to cover the park itself, while one or more macro cells could provide overall coverage of the entire area. Figure 8-11 shows what such an arrangement might look like.

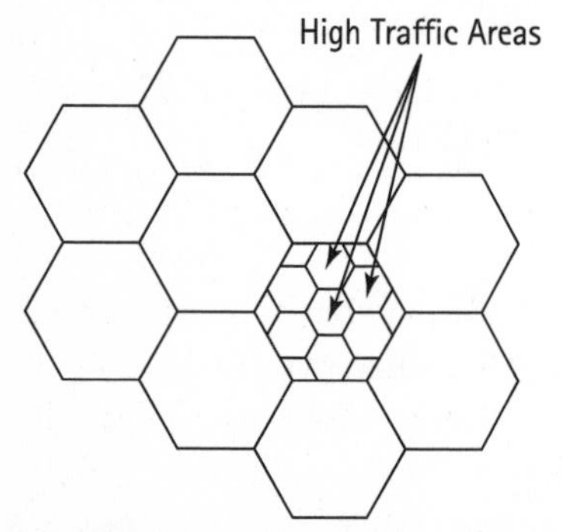

Figure 8-11: Using HCS to cover a small area with a high cellular-user density can be more efficient than using cell splitting.

HCS is effective because TDMA phones are smart. In other words, they can be programmed to choose the appropriate type of cell for a given situation. In a business park, for example, with phones programmed to use micro or pico cells when available, most of the wireless activity in the park would go through the nearest available micro cell, instead of tying up a channel in the sole macro cell in the vicinity.

TDMA phones also contain special algorithms that can determine when the phone is moving rapidly through an area of micro or pico cells. In such cases, the phone may determine that it is most efficient to try to communicate with the nearest macro cell, instead of the smaller cells. Doing so can greatly reduce the number of times the phone is handed off from cell to cell, thereby saving on processing power and reducing the work for the system as a whole.

Pico cells also have a lot of potential for indoor use, say in a subway system. By bringing these small cells into buildings and other locations, gaps in coverage can be filled, and carriers can earn extra revenue by enabling users to be online more often.

TDMA devices can support a wide range of digital features, limited primarily by what the network operator cares to provide and what the user is willing to pay for.

Currently, three main wireless standards compete with one another for the United States market: TDMA, CDMA, and GSM. However, TDMA is on its way out. As part of a $10 billion deal between AT&T Wireless and NTT DoCoMo, AT&T Wireless will be phasing out TDMA support over the next year or so and switching over to GSM. Cingular Wireless, the other service offering TDMA support in the United States, is expected to follow suit. While this is bad news for TDMA fans, or people using TDMA-based phones, it is good news in general. Reducing the number of wireless technologies in use increases the odds of getting good nationwide coverage for the technologies that remain (CDMA and GSM primarily).

Summary

GSM is a sophisticated, feature-rich technology that is exceedingly popular around the world, with over half a billion users and growing. People often discuss 2G wireless technology as if it was a single technology. Instead, as this chapter has shown, 2G is a convenient handle for a number of incompatible technologies. For example, due to its use of three incompatible 2G standards, the United States is behind the rest of the world in its development of usable, practical 2G wireless services.

2G technologies take advantage of the conceptual foundation laid by 1G technologies, particularly the idea of a cellular network. In fact, some (such as CDPD) are built directly on the existing 1G infrastructure. Some 2G technologies, CDMA for example, required the deployment of an entirely new, all-digital infrastructure.

Chapter 9

2.5G Techs and Specs

In This Chapter

This chapter covers the following main topics:

- ✓ Understanding 2.5G concepts
- ✓ From GSM and TDMA to IMT-2000
- ✓ HSCSD
- ✓ GPRS
- ✓ EDGE
- ✓ From CDMA to IMT-2000
- ✓ IS-95B

A World of Heightened Expectations

In terms of technological development, technophiles tend to see only "one right way," which involves pushing our technologies as far and as fast as we can. We are guilty of this attitude, with all our acquaintances tired of hearing us complain about how the U.S. lags far behind the rest of the world in wireless. After all, several designs for 3G wireless technologies exist. And they even (sort of) work. So why the heck can't we buy a 3G phone? Why can't we get high-speed Internet access on a PDA? If we can put a man on the moon, why can't we get 100 percent wireless coverage across the country?

When we look at the question logically, the answer is clear: cost. There are limits on what we as a society — even as a world-spanning civilization — can spend on technology (even really cool technology such as 3G data terminals).

When talking about 2.5G and 3G technologies, the common terminology for a mobile phone or other wireless device is *data terminal.* You will see this term in this chapter and the next.

In other words, economic reality sometimes conflicts with our techno-fantasies. That's why you are reading a chapter on 2.5G wireless technologies. We know how to go from 2G technology to 3G technology. There is even some international agreement on how to bring all of the existing

2G networks together into a globally-compatible 3G network. But no one can come up with the money to get there all in one big leap.

The International Telecommunication Union (ITU) has the road to 3G all mapped out as part of their IMT-2000 plan. You can see it for yourself at the ITU Web site, `www.itu.int/imt/`. You'll find specs, charts, white papers, and seminars — virtually anything you may need to know about ITU's vision for 3G wireless.

Preparing for 3G

Over the last few years, the phone companies and other groups interested in owning a piece of the 3G radio spectrum have spent tens of billions of dollars buying licenses auctioned off by national governments. Those licenses allow the holder to provide wireless services in a specific frequency band and throughout a certain geographical area.

While the governments that ran the auctions made a pile of money, the groups that won the auctions found themselves in a bit of a dilemma. A few of them borrowed huge amounts of money to buy licenses, only to find that they couldn't make their payments and went bankrupt. Others haven't gone bankrupt, but between their debt service and reduced revenues due to the current economic slowdown, they don't have the money to build their 3G network.

The first part of 2001 has seen a number of setbacks for 3G technologies. These setbacks appear to be the standard kind of problems experienced when any complex new technology is developed, so they should be temporary. Even so, the schedules for deploying 3G technologies keep slipping; it is now unclear how many more years it will take before they are widely available.

Therefore, we need a short-term solution that:

- ✓ Gives users faster service and cooler features.
- ✓ Gives the wireless industry something new to sell.
- ✓ Can be deployed within the next few years.
- ✓ Doesn't cost untold billions to build.

The name of this solution is 2.5G.

Understanding 2.5G Concepts

2.5G technologies are technologies built onto existing wireless networks. They provide some of the benefits promised with 3G, such as faster data transfer, "always-on" service, and the ability to browse the Web. However, because they are built on the existing networks, they cost far less than 3G systems, which require the creation of entirely new physical networks.

Beyond that, the 2.5G technologies are pretty diverse. Some use TDMA and others use CDMA. Most are packet-switched; one is still circuit-switched. They can move data at anywhere from under 30 Kbps to over 300 Kbps.

In short, what defines a 2.5G technology isn't the specific technology itself; rather, the 2.5G technology takes us part of the way from 2G to 3G by giving us the features and capabilities we want most (in particular, higher data rates) and doing so in a way that is affordable.

From GSM and TDMA to IMT-2000

GSM and TDMA (IS-136) are alike in some important ways. Most importantly, both use time-division multiple access when dividing the available channels to support multiple users. Thus, they can follow similar paths toward 3G, even in those cases where the IS-136 network operator will eventually implement TDMA-136 instead of WCDMA as his 3G technology.

Of the three 2.5G technologies discussed in this section, two (GPRS and EDGE) can be applied to either GSM or IS-136 networks. The third, HSCSD, is a GSM-only technology (and it isn't even really on the path to 3G). Since GSM is by far the most popular of the 2G standards, and because HSCSD, GPRS, and EDGE were all initially designed to work on GSM networks, we will treat them as if they were working with a GSM network in the paragraphs that follow.

HSCSD

HSCSD stands for *High Speed Circuit Switched Data*. This technology is an enhancement to the standard GSM data services. HSCSD enables customers to get higher data rates on existing GSM networks through the use of a special PC card or HSCSD data terminal. As of December 2001, HSCSD service was available to more than 90 million GSM customers worldwide.

Table 9-1 lists the key characteristics of HSCSD.

HSCSD applies only to GSM, not TDMA (IS-136), networks.

TABLE 9-1 **HSCSD at a Glance**

Characteristic	Value
Base Network	GSM
Initial Deployment	2000

Continued

TABLE 9-1 **HSCSD at a Glance** *(Continued)*

Characteristic	Value
Wireless Generation	2.5G
Signal Type	Digital
Spectrum Division	FDMA/TDMA
Switching	Circuit-switched
Data Rate	28.8 Kbps typical

Due to its relatively high data rates in a mobile phone environment, HSCSD is considered by most to be a 2.5G technology. Even so, you should think of it as a path independent from the one that takes GSM to IMT-2000. The reason is simple: Adding HSCSD to an existing GSM network is not a necessary part of the transition to IMT-2000. That said, HSCSD is a bona fide 2.5G technology; thus, it is covered here. Think of HSCSD as an easy-to-implement shortcut to getting high-speed data service on GSM networks while waiting for GPRS to arrive.

Figure 9-1 shows where HSCSD sits along the planned evolutionary path of GSM.

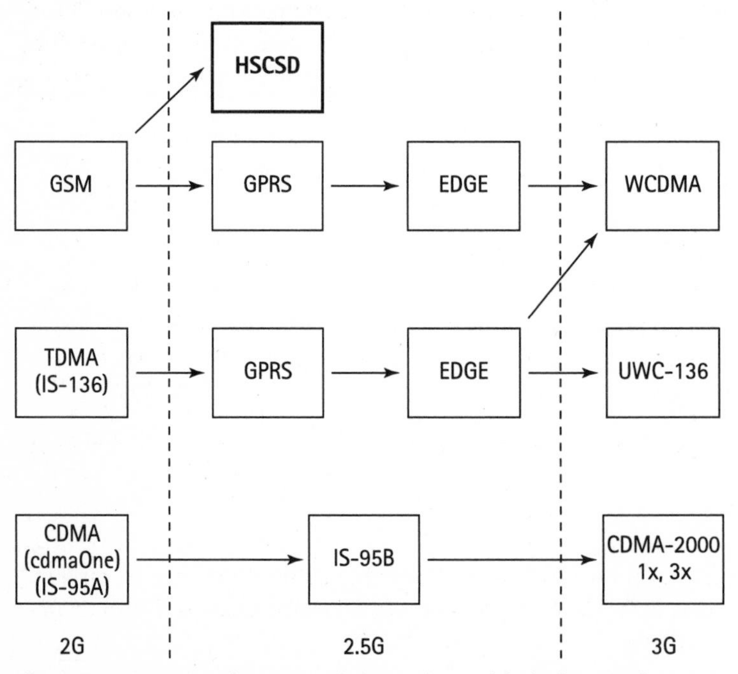

Figure 9-1: HSCSD is a short-term solution to the need for higher-speed data transfers on GSM networks.

HOW HSCSD WORKS

As you recall, GSM effectively divides each channel into eight time slots, allocating one slot to each of eight data terminals, thereby allowing eight simultaneous users per channel. Each of those users gets 9.6 Kbps of bandwidth on that channel. HSCSD gets its higher data rates by allowing one data terminal to use more than one time slot, effectively multiplying the network's standard data rate by the number of time slots the data terminal is using. In most cases, adding HSCSD support to an existing GSM network requires only a software upgrade. However, it does require special HSCSD data terminals.

Initial HSCSD installations enable a data terminal to use three time slots, giving it a data rate of 28.8 Kbps, instead of the usual 9.6 Kbps. Some GSM networks also offer 14.4 Kbps data rates per time slot. So far, those networks are also offering HSCSD at 28.8 Kbps by allowing HSCSD data terminals to use two time slots.

Depending on how the service is implemented by a particular carrier, it is possible that an HSCSD data terminal will not be able to transmit or receive at its full rated speed. During times of heavy network traffic, there may not be enough available time slots for the data terminal to use. In such cases, HSCSD data terminals are designed to work with the available network resources, shifting to two time slots at a time, or even one, instead of the usual three.

An additional factor for HSCSD data terminals is that, in general, high-speed data transfer is far more important when the data is moving from the network to the data terminal — that is, when the user is receiving data. Some HSCSD data terminals can be configured with up to three time slots allocated for receiving data from the network, while a single time slot is reserved for sending data to the network.

Initially, handheld HSCSD data terminals were unavailable; today, there are numerous choices, many of which can also serve as wireless modems for a laptop or other computer.

Due to the way HSCSD uses multiple time slots, it can increase the load on the network, as well as cost more to use than standard GSM service. In addition, it requires specialized hardware, and is only available in limited geographic areas. And it will be superseded by GPRS within a few years. Make sure you really need (and have access to) the higher data rates HSCSD offers before choosing this technology.

BENEFITS OF HSCSD

HSCSD offers the following advantages over GPRS:

- ✓ It is more widely available right now (that is, as of summer 2001). More than 90 million users currently have access to HSCSD. The number of users with access to GPRS today is somewhere around 1 million.

- ✓ Since HSCSD is circuit-switched, the available bandwidth is consistent over the course of a connection, making it more suitable for applications such as video conferencing and multimedia.
- ✓ Network operators and users are already familiar with billing models for circuit-switched connections. HSCSD is billed by connect time — just like the rest of the GSM network it is installed on — making it simple to update the billing system and reducing the number of billing-related support calls from subscribers.

GPRS

General Packet Radio Services, most commonly known as *GPRS*, is an evolutionary link between GSM and 3G wireless services. GPRS works on the existing GSM network (with the appropriate transceivers added and with a GPRS-ready data terminal in hand) and provides much higher data-transfer rates than the existing network — theoretically, up to 171 Kbps, or three times faster than the current GSM networks.

GPRS trials are underway in Europe, with hopes that the first general GPRS deployments will occur before the end of 2002. Some limited systems are already in operation in Asian countries, including Singapore, Taiwan, and China.

Table 9-2 summarizes the basic information about GPRS.

TABLE 9-2 GPRS at a Glance

Characteristic	Value
Base Network	GSM
Initial Deployment	2000
Wireless Generation	2.5G
Signal Type	Digital
Spectrum Division	FDMA/TDMA
Switching	Packet-switched
Data Rate	171.2 Kbps maximum

Figure 9-2 shows where GPRS fits into the overall evolution of GSM from its current 2G state to full 3G-compliance, as defined by the ITU in its overall plan for IMT-2000, an integrated plan for interoperability of 3G wireless networks. IMT-2000 is also referred to as UMTS, FPLMTS, or J-FPLMTS.

In addition to greater speed, GPRS offers two major benefits to users. First, whenever a GPRS data terminal is on and within the coverage area, it is connected to the GPRS network. That means much quicker response time. There is no need to wait for the data terminal to dial into the network and establish a circuit before sending a chunk of data.

The second major benefit is the availability of numerous capabilities that aren't practical on the existing GSM networks or are too cumbersome to be of much use. Two examples are Web browsing and the efficient use of WAP. With GPRS, data terminal users will have full access to the entirety of the Internet for the first time, although they will still be limited in terms of what they can do. For example:

- ✓ GPRS isn't fast enough to display full-motion video.
- ✓ Phone displays are too small to get the full effect of most Web pages (although Smartphones, with their larger displays and clever browser designs, may come close).
- ✓ The user interface on a mobile phone is too limiting for many Web applications (here again, Smartphones have an edge, with their touch-sensitive screens, handwriting recognition software, and virtual keyboards).

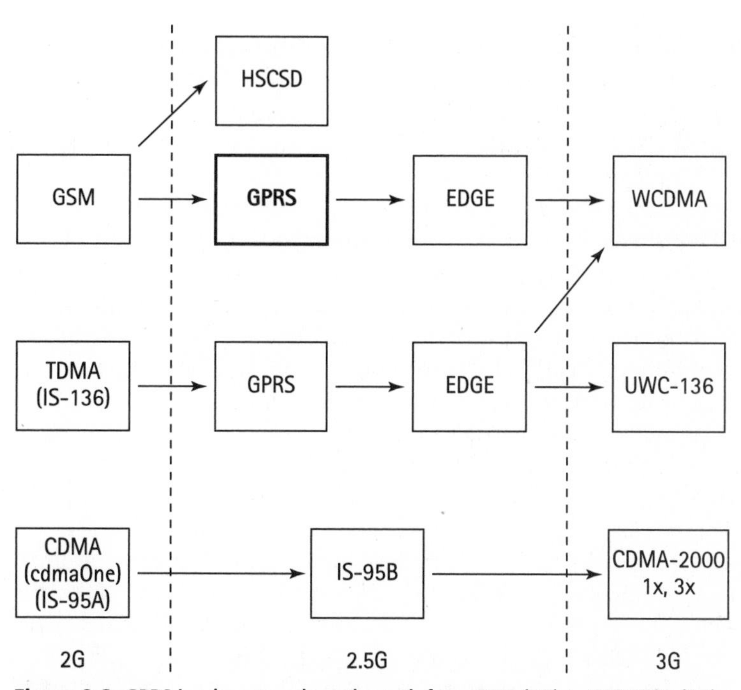

Figure 9-2: GPRS is a key stop along the path from GSM (2G) to IMT-2000 (3G).

You can get the latest news on GPRS at the MobileGPRS.com Web site: `www.mobilegprs.com`. The site features news, discussions, various papers and reports, developer information, and even a list of current GPRS trial and deployment contracts. The site is part of a vaster site run by Mobile Lifestreams. The mega-site covers more than 90 nonvoice mobile data communication technologies, dozens of which will be of interest to anyone reading this book.

HOW GPRS WORKS

In contrast to GSM, which is a *circuit-switched* technology, GPRS is *packet-switched.*

GPRS PACKET-SWITCHING VERSUS GSM CIRCUIT-SWITCHING If you read Chapter 8, you already know that the term *circuit-switched* means that there is, in effect, a dedicated circuit from data terminal to base station for the duration of a call. You will also remember that the term packet-switched means the data is chopped into little packets, each transmitted individually. Furthermore, when not actively transmitting a packet, *packet-switched* data terminals don't tie up any circuits or channels or time slots. GPRS is a packet-switched technology that runs on the existing GSM infrastructure. Therefore, the GPRS packets must somehow fit into the FDMA/TDMA design of GSM. The FDMA part is easy. Since GSM divides the available bandwidth into channels, GPRS must work within those channels.But what about the TDMA aspect of GSM? In effect, GSM divides each channel into eight time slots and assigns each channel to a different data terminal. Each data terminal transmits or receives only within its own time slot.

To implement GPRS, the system needs to take some of these time slots and use them for data packets instead of GSM data. With the appropriate hardware on each end (users *and* network providers must use new hardware to work with GPRS), the system can work with time slots containing GSM data as it always has, while working with time slots containing GPRS packets the way they need to be handled.

TAKING ADVANTAGE OF THE GPRS NETWORK If GPRS simply meant replacing the GSM data in a time slot with GPRS packet-switched data, there would be real advantages over GSM as it is today. Multiple GPRS data terminals could share each time slot, since they only use the slot when they are actively transmitting data, which would increase overall network capacity. Plus, the GPRS data terminals could be continuously connected, providing "always-on" connectivity. But there is also something you can do with GPRS data packets to greatly increase the speed at which data travels across the network. If you move voice data too fast, you need a system to store it and play it back at the proper speed and in proper time relationship with the packets the came before and after. Since GPRS just carries data, the faster that data crosses the network, the better. There is no such thing as moving data too fast. So GPRS data terminals can transmit and receive in multiple time slots. Say, for example, you are browsing a Web site with your GPRS data terminal, and there isn't much traffic on the wireless network. The network could conceivably send packets to your data terminal using each of the eight time slots of the channel your data terminal is on. Your data terminal could conceivably receive (and also transmit) eight times as much data in a given time period as it would if the system were restricted to one time slot per GPRS data terminal, as in the circuit-switched GSM approach.

Of course, if other data terminals, such as GSM or GPRS, are using the same channel as your GPRS data terminal, the data terminal won't be able to achieve maximum speed, since some of the time slots will be in use at least part of the time. Thus, the data rate a GPRS data terminal will achieve is highly dependent on outside factors — namely, the number of other data terminals using the network at the same time.

You've seen the technical benefits of GPRS packet switching, the way it allows the network to carry more users and each user to get (in general) higher-speed data connections. But GPRS packet switching also offers financial benefits. For example, imagine that you have a sibling who is still using GSM, while you have just moved up to GPRS. When your sibling's GSM data terminal sends or receives data across the network, it is directed to use a specific channel and one of eight time slots within that channel. From that point on, your sibling's GSM data terminal retains control of that slot within that channel for the duration of the call. Even when the data terminal is not sending or receiving data, it ties up one slot on one channel.

The preceding explanation is an oversimplification of the way GSM actually handles channels and time slots. It is, however, accurate enough for the purposes of this example. If you want a more detailed explanation, read "How GSM Works" in Chapter 8.

If your sibling spends ten minutes reading the article without disconnecting the GSM data terminal from the network, he or she will be charged for using 10 minutes of GSM service. This is only fair. Since no one else can use the same channel and time slot your sibling's data terminal is using, he needs to pay for tying up that resource — even if the data terminal isn't actually transferring data.

Expert Advice: Understanding the Fundamental Difference between GPRS and GSM

Three key characteristics of GPRS combine to make data terminals that use it (and the technologies that will follow it) fundamentally different from data terminals that use GSM. Packet-switching, continuous connectivity to the network, and reasonably high data rates make GPRS data terminals more like personal computers permanently connected to the Internet than phones occasionally dialing into data services. Aside from the fact that GPRS data terminals can support more interesting and capable applications than GSM data terminals, there is a basic difference in perception and the way you want to use a fast, always-on Internet-connected data terminal.

For example, when you may have first brought a PC with a modem into the house, connecting to the network was slow, and when the connection was finally made, almost every activity took a long time. In short, being on the computer may not have been worth the effort.

Switching to a cable modem, however, gives you a packet-switched, continuously connected PC with a reasonably fast data rate. Suddenly, your computer becomes much more interesting and useful. You probably use it at least a couple more hours every day. It is possible a similar phenomenon will occur when GSM users move to GPRS. Assuming that the phones themselves are easy to use, and that it is easy to work with the data and applications that become available, people are likely to spend a lot more time actively using their GPRS data terminals. Smartphones, with their more powerful user interfaces, larger displays, and greater processing power, are likely to come into their own at this point.

Now, suppose you use your GPRS data terminal to download the same article from the same WAP site as your GSM-using sibling; then you spend ten minutes reading it (without disconnecting from the network — remember that GPRS data terminals are always connected). You will be charged only for the amount of data you downloaded. This is also fair. Since your GPRS data terminal only uses time slots to transmit and receive data, then frees them for other data terminals to use, others can use the same channel and time slot you were using. You shouldn't have to pay for the time during which your GPRS data terminal is just sitting there, monitoring the packets that come across the network.

The other key to understanding how GPRS works is that voice calls and data calls travel different paths through a GPRS network. Separating the voice and data allows the voice to continue to travel over the existing GSM infrastructure, while the new GPRS systems can be optimized for moving data. The GPRS specification defines three types of mobile data terminal. Class A terminals support simultaneous voice and packet-switched transmissions. Class B terminals support voice and packet-switched transmissions, but not simultaneously. Class C terminals support only packet-switched transmissions.

To add GPRS support to existing GSM networks, the service provider must add two new components:

- ✓ **Gateway GPRS Support Node (GGSN):** This component provides the connection between the GPRS network and IP-based or X.25-based external networks. It communicates with the existing GSM network through the SGSN.
- ✓ **Serving GPRS Support Node (SGSN):** This component provides authentication and mobility management for GPRS users. It communicates with the existing GSM network's Mobile Services Switching Center (MSSC) and the Home Location Register (HLR), and provides a connection from the GGSN to the existing GSM network. It functions similarly to the MSC.

Figure 9-3 shows these two new components.

GPRS IMPLEMENTATION DETAILS

Now that you understand the major components that add GPRS support to GSM networks, it is time to look at some implementation details. These include air-interface changes and connections to external-packet networks.

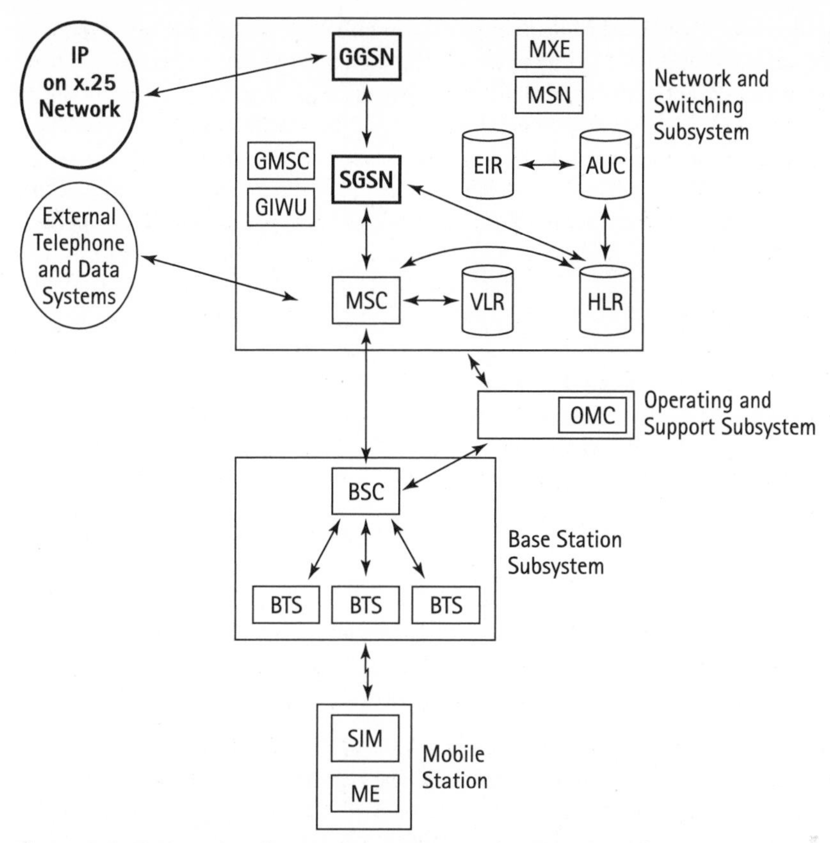

Figure 9-3: GPRS requires that two new components (GGSN and SGSN) be added to an existing GSM network.

AIR-INTERFACE CHANGES In order for GPRS data packets to travel on the GSM network infrastructure, changes to the air interface and the systems that serve it are necessary. To begin, any Base Transceiver Station (BTS) that supports GPRS must assign one or more channels to the task. Such channels are called *Packet Data Channels (PDCHs).* Logically, each PDCH is divided into three channels:

- ✓ **Packet Data Traffic Channel (PDTCH):** This channel carries the encoded user data for each packet.
- ✓ **Packet Broadcast Control Channel (PBCCH):** This channel carries general system information.
- ✓ **Packet Common Control Channel (PCCCH):** This channel carries information on the imminent start of a packet transmission. It also notifies data terminals of the availability of additional time slots to transmit on.

Packet data traveling on the PDCH gets interleaved across four time slots to create a packet with 456 bits of coded data. GPRS uses four different coding schemes, each optimized for data transmission and each capable of moving data across the network faster than with GSM.

CONNECTING TO EXTERNAL-PACKET NETWORKS Connections to external IP-based or X.25-based networks are the province of the GGSN. The GGSN provides an address for each data terminal that is appropriate to the external network the GGSN is connected to. To acquire such an address, a data terminal (through the SGSN) requests a Packet Data Protocol (PDP) context between the SGSN and the GGSN. The GGSN then transfers data between the PDP and the external networks.

For more in-depth discussion about GPRS, visit MobileGPRS.com (`www.mobilegprs.com`) and GSMWorld.com (`www.gsmworld.com`).

EDGE

Depending on your source, EDGE stands for *Enhanced Data rates for GSM and TDMA/136 Evolution*, *Enhanced Data rates for Global Evolution*, or *Enhanced Data rates for GSM Evolution*. Acronyms aside, EDGE refers to the last planned stop along the path from GSM to IMT-2000, the ITU's international 3G standard.

Table 9-3 summarizes some basic information about EDGE.

TABLE 9-3 EDGE at a Glance

Characteristic	Value
Base Network	GPRS
Initial Deployment	2001 (?)
Wireless Generation	2.5G
Signal Type	Digital
Spectrum Division	FDMA/TDMA
Switching	Packet-switched
Data Rate	Over 384 Kbps maximum

Figure 9-4 shows the place EDGE occupies on the overall map of the road to IMT-2000 (3G).

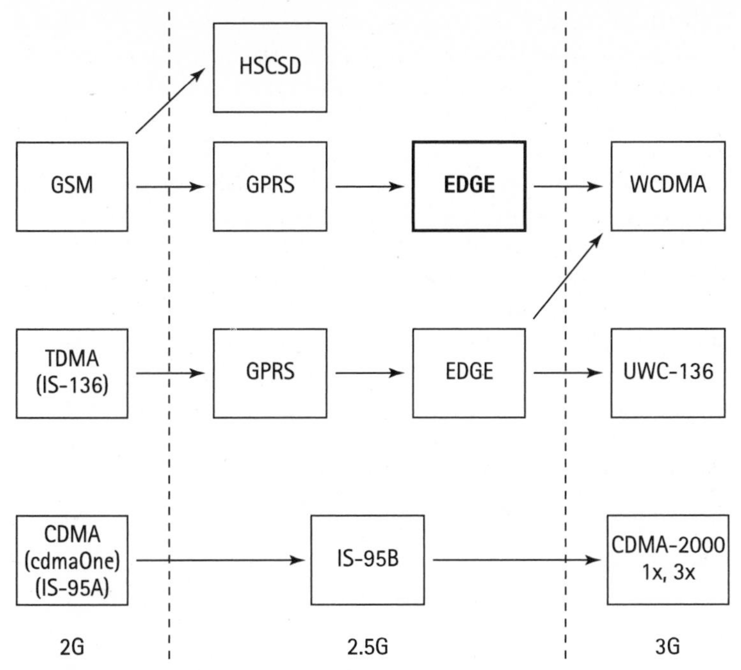

Figure 9-4: EDGE is the last planned stop along the path from GSM (2G) to IMT-2000 (3G).

EDGE builds on the capabilities of the GPRS air interface and network hardware to improve the performance and efficiency of the network. EDGE uses new coding schemes and adaptive modulation (see the following section, "How EDGE Works," for more details) to move data faster and to adapt to the quality of the air interface at any instant, allowing the overall communication channel to function as efficiently as possible given environmental conditions.

Since EDGE was originally designed to work on GSM networks, it can freely share those networks with GSM data traffic. On TDMA networks, some of the benefits of EDGE are lost, as EDGE channels need to be segregated from regular TDMA channels. Even within the EDGE channels, some time slots must remain unused to reduce the possibility of *co-channel interference*, which limits the benefits of EDGE in TDMA networks.

HOW EDGE WORKS

EDGE institutes a standard set of improvements to GPRS. It is particularly well adapted to improving GPRS on GSM networks, as it is designed to use the same channel and time slot structure as GSM. The improvements delivered by EDGE enable networks to move data faster and to more efficiently use the available radio spectrum. These improvements are the following:

- ✓ Adaptive radio link
- ✓ 8-PSK modulation

ADAPTIVE RADIO LINK In GPRS systems, the modulation and coding of the packets remain the same, no matter the quality of the signals moving through the channel. The system is designed to deliver data the same way, and at the same rate, whenever the quality of the overall channel meets the minimum requirements.

EDGE is designed to detect the quality of the channel and deliver data as fast as possible, given channel conditions at any given time. If the quality of the channel is high (perhaps because the data terminal is nearby and weather conditions are optimal), EDGE will automatically deliver data faster.To do this, EDGE doesn't actually change the rate at which signals travel through the network. Instead, as the quality of the channel improves, EDGE uses higher-level modulation (that is, fits more data into a given number of bits) and reduces coding (that is, provides less protection for the data) to move more data across the channel.

8-PSK MODULATION 8-PSK (8-Phase Shift Keying) modulation is a technique for modulating a signal with three bits of data for each modulated symbol. It provides approximately twice the efficiency as the GSMK modulation technique used by GSM and GPRS.

ADDING EDGE TO A GPRS NETWORK

Adding EDGE to a network that has already been upgraded for GPRS is relatively simple. An EDGE transceiver must be added to each cell, and the base station software updated to work with that transceiver.

From CDMA to IMT-2000

Before you can understand the evolution of CDMA systems from 2G to 3G, you need to be clear on some terminology. *CDMA* (or *Code Division Multiple Access)* is a technology that can be applied in various ways, to various networks. For example, GSM is evolving to a CDMA-based 3G version known as WCDMA.

Now consider the term *cdmaOne*. cdmaOne describes a complete wireless system that includes the CDMA/IS-95 air interface, the ANSI-41 network standard for switch interconnection, and several other standards to form a complete wireless system.

cdmaOne is a brand name backed by the companies that make up the CDMA Development Group (CDG). This organization promotes cdmaOne and has created its own 3G standard, called cdma2000.

cdma2000 is designed to meet the requirements of the ITU's IMT-2000, and it provides an easier upgrade path for cdmaOne operators than would migrating to WCDMA. It is also considered compatible with the existing TDMA and GSM 2G standards, making it possible for operators to migrate to cdma2000 for their 3G implementation, instead of WCDMA or UWC-136. cdma2000 has been accepted by the ITU as one of the IMT-2000 3G technologies.

Perhaps the best way to look at this particular situation is that the operators of cdmaOne networks have arranged for their own separate path to the 3G IMT-2000 standard, which would certainly make sense for those who've purchased cdmaOne-based equipment. These users have also defined a single 2.5G intermediate point between cdmaOne (2G) and cdma2000 (3G). This intermediate point is known as *IS-95B*.

IS-95B

IS-95B is the 2.5G version of cdmaOne. Figure 9-5 shows where IS-95B fits into the overall 2G to 3G transition path for the industry.

Functionally, the most obvious difference between the original cdmaOne (IS-95A) and IS-95B is speed. While the original version has a maximum data rate of 14.4 Kbps, IS-95B can move data at a sustained rate of 65 Kbps. However, this is only possible when the IS-95B data terminal is connected to an IS-95B or later network. When connected to an earlier cdmaOne network, an IS-95B data terminal will communicate at 14.4 Kbps. IS-95B also implements numerous other features and improvements.

For the latest information on CDMA, visit the CDMA Development Group Web site at `www.cdg.org/`.

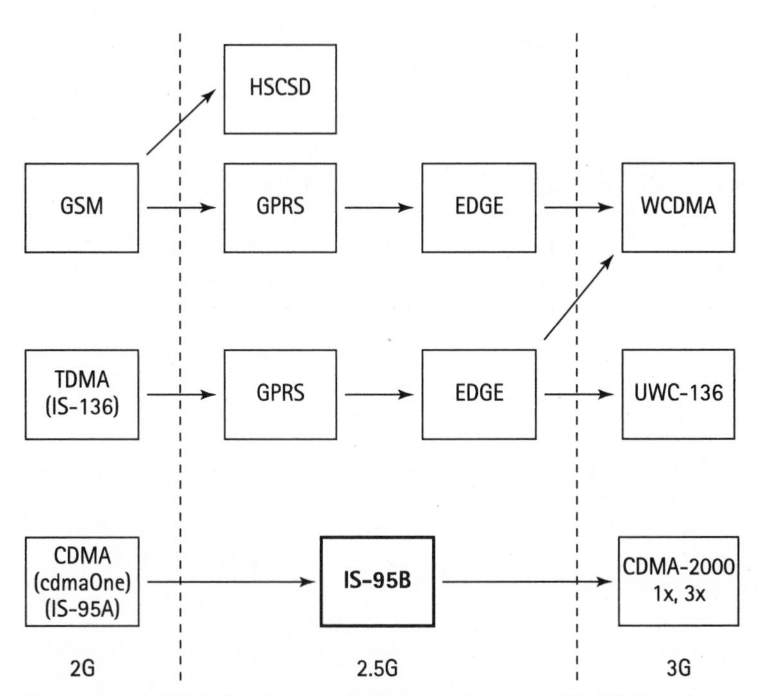

Figure 9-5: IS-95B is the midway point between the current 2G CDMA technology (cdmaOne) and the 3G destination cdma2000.

Table 9-4 lists the main features of IS-95B.

TABLE 9-4 **IS-95B at a Glance**

Characteristic	Value
Base Network	CDMA
Initial Deployment	1999
Wireless Generation	2.5G
Signal Type	Digital
Spectrum Division	CDMA
Switching	Packet-switched
Data Rate	14.4 Kbps on IS-95A networks
	65 Kbps on IS-95B and later networks

IS-95B networks have been in service since September 1999, when the first commercial service was launched by Korea Telecom Freetel, Inc.

How IS-95B Works

Implementing IS-95B requires the addition of hardware and software to the system's Base Station Controllers.

Like the original version of cdmaOne, IS-95B is IP-based. This means that an IS-95B network can use standard routers and other equipment that are normally used for Internet connectivity. This provides real cost savings and ease of maintenance for network operators. In addition, IS-95B networks and data terminals are forward- and backward-compatible with existing and upcoming networks. This means that it won't be necessary to replace terminals every time a more capable version of the network goes live, although it will be necessary to replace terminals to take full advantage of the new networks.

In addition to higher speed, IS-95B provides a number of capabilities, including the following:

- ✓ A new soft handoff algorithm
- ✓ A new interfrequency hard handoff capability
- ✓ Search enhancements
- ✓ Enhanced power control

These incremental improvements enable IS-95B systems to support some nice new features, including the following:

✓ Calling name presentation (caller ID)

✓ Priority access channel assignment (ensures access to a usable channel in an emergency)

✓ Power up function (increases data terminal transmit power for a short time to ensure that it is detected by all nearby base stations)

Summary

From the very terms used to describe them, you can tell that the 2.5G technologies covered in this chapter were afterthoughts. Everything was moving along in a nice progression — 1G, 2G, 3G — and then WHAM!

Reality hit. The need for tens of billions of dollars of investment, on top of the tens of billions of dollars that had already been spent on 3G-spectrum auctions, put a damper on the dream of high-speed, broadband, full-color, full-motion everything available all the time through your mobile phone — at least for the time being. We still think most of the hype will come true. It will just take years longer than we expected it to.

The end of the dream of 3G super phones arriving on the scene right away left the industry with a big problem. The answer was 2.5G, a set of technologies and fixes and tricks and hacks that can deliver at least some of the 3G promise for a whole lot less money. At this time, 2.5G networks are spreading, and 3G seems to be receding into the distance. Therefore, since we are stuck with 2.5G technologies for a while, here are some thoughts you should take away from this chapter:

✓ Thanks to some key similarities, GSM and TDMA (IS-136) systems can share much the same evolutionary path, meaning that GPRS and EDGE can work on both.

✓ HSCSD is a way to get high-speed data on GSM networks today — if you're in a coverage area, don't mind buying a new data terminal, and don't mind the thought that GPRS will almost surely replace HSCSD within a year or so.

✓ GPRS is beginning to spread across Europe and Asia, bringing higher-speed data plus the always-on benefits of packet switching to GSM.

✓ When EDGE arrives, it will make GPRS even faster and more efficient, with little additional cost.

✓ IS-95B is bringing higher speed and some useful additional features to systems built around cdmaOne.

Chapter 10

3G Techs and Specs

In This Chapter

This chapter takes you beyond the state-of-the-art and into the world of 3G wireless systems. Although limited tests have been conducted or are underway around the world, the arrival of 3G is at least few years away — perhaps more than that, given the current economic climate.

However, when they do arrive, 3G technologies and systems should live up to all the hype we've been hearing about the "wireless revolution." With the (planned) ability to transfer data at rates of around 2 Mbps, 3G terminals will be able to perform in ways that simply aren't feasible right now. Add to this the fact that 3G systems are designed as always-on, Internet-compatible, all-purpose multimedia devices, and you have science fiction coming into existence in the palm of your hand.

This chapter discusses 3G technologies, with a focus on the three 3G technologies that are currently approved by the International Telecommunications Union (ITU) as part of its international plan for 3G systems: WCDMA, UWC-136, and cdma2000.

In addition, this chapter covers some related technologies. TD-SCDMA, for example, is an alternative technology that was proposed, but not approved, as part of the final 3G standard. Another technology, known as *smart antenna*, is used to increase the speed and performance of virtually any existing air interface, perhaps boosting speeds enough to make an earlier technology competitive with 3G, at least where raw speed is concerned.

Given the state of development of 3G, the economic climate at the time this book was written, and the uncertainty about exactly what users want when they think of 3G technology, it is highly likely that much of the information in this chapter will turn out to be obsolete. Today, a service or product's market image, perception, and positioning may be as important to eventual success in the 3G world as specific technologies. Everyone is scrambling to design affordable systems that satisfy the needs of their constituencies (GSM or cdmaOne network operators, for example), and that meet, or at least come acceptably close to, the IMT-2000 requirements of the ITU. Furthermore, the proposed designs will have to be realistic — solutions that can actually be built and deployed successfully.

Despite some highly publicized 3G pilot programs, it is still early in the 3G life cycle, and many questions remain unanswered. The final 3G systems that we end up using in the year 2005 will certainly differ significantly (at least in terms of technical details) from the technology described in this chapter.

This chapter covers the following topics:

- ✓ Understanding 3G concepts
- ✓ IMT-2000 3G technologies
- ✓ WCDMA
- ✓ UWC-136
- ✓ cdma2000
- ✓ Related technologies
- ✓ TD-SCDMA
- ✓ Smart antenna technology

Understanding 3G Concepts

It is important to remember that 3G technology is not a single technology. At one time, the ITU attempted to put a single *Radio Transmission Technology (RTT)* (another term for air interface) into worldwide use. That effort failed.

The ITU's current plan for 3G, known as *International Mobile Telecommunications-2000 (IMT-2000)* shows three approved RTTs: WCDMA, UWC-136, and cdma2000. In addition, there are rumblings that TD-SCDMA, a technology supported by Chinese developers, will soon be added to the list of approved RTTs. A discussion about TD-SCDMA appears in the "Related Technologies" section of this chapter.

The ITU Web site provides extensive details on IMT-2000, including a roadmap, video clips, and links to related sites. The address is: `www.itu.int`.

While IMT-2000 has evolved to allow multiple RTTs, it is still meant to bring together competing standards, the ultimate goal being to meet the needs of the customer within the limits allowed by regulation, markets, and the business requirements of the companies actually delivering the services.

Some steps in the direction of a common standard include the development of common billing and user profiles across the different technologies, as well the agreement to provide certain common capabilities, such as bandwidth on demand and mail storage and forwarding.

Finding the Spectrum Needed for 3G

The overall frequency bands allocated for IMT-2000 use by the World Radio Congress are 1885–2050 MHz and 2110–2200 MHz.

A portion of the radio spectrum that will be used for 3G systems in the United States is at 1895–1910 MHz and 1975–1990 MHz. Licenses to much of that spectrum were originally purchased in 1998 by NextWave Telecom, a now-bankrupt company. Among the markets covered by these licenses are New York City, Detroit, Denver, and Seattle. Unfortunately, a significant legal battle now revolves around those licenses.

After NextWave filed for bankruptcy protection, the United States Federal Communications Commission (FCC), repossessed those licenses, then resold them in early 2001 for nearly $16 billion to several companies, including Verizon Wireless, VoiceStream Wireless, and Alaska Native Wireless.

In June 2001, an appeals court ruled that the FCC had violated U.S. bankruptcy laws by repossessing the licenses. In July 2001, the three companies that purchased the repossessed spectrum licences requested that federal regulators investigate NextWave to see if the company actually meets eligibility requirements to hold wireless licenses.

This story illustrates how desperate companies are to find the spectrum needed to deliver 3G services. And through it all, the people of New York City, Detroit, Denver, and Seattle wait patiently for improved wireless services — if only the licensed spectrum was put to good use.

Despite problems such as lack of bandwidth and even lack of availability of compatible handsets, several companies remain committed to getting a minimal 3G network up and running, possibly as early as 2002. Verizon Wireless announced on August 1, 2001, that it had upgraded to 3G a portion of its network running within a 100-square-mile area of New York and New Jersey. Likewise, Sprint PCS planned to have a pilot system up and running in one city by September 2001, with a full, nationwide commercial launch planned for June 2002. Overseas, NTT DoCoMo was planning a launch for October 2001.

Meanwhile, at the other end of the line, Nokia continues to promise to deliver 3G handsets by the end of 2001. Their commitment to this goal is important, since manufacturers have consistently experienced problems producing 3G handsets that are ready for a mass audience. Hopefully, by the time you read these words, the major problems will have been ironed out, and 3G systems will be spreading across the globe.

IMT-2000 also has a space-based component; eventually, it should allow satellite networks to supplement the terrestrial wireless networks in areas where no terrestrial service is available. The space-based component of IMT-2000 is even less developed than the terrestrial component. But the implementation of this component is several years from now.

Key Features of IMT-2000-Compatible Systems

All IMT-2000-compatible systems will have several key features, including the following:

- ✓ Support for a range of mobile and fixed terminals
- ✓ A high degree of design commonality
- ✓ Small terminals
- ✓ Compatibility with other services within IMT-2000 and with wired networks (the worldwide wired phone system in particular)
- ✓ Worldwide roaming
- ✓ High quality
- ✓ Support for multimedia applications and a wide range of services

If IMT-2000 can deliver this comprehensive list of features, we should be carrying some pretty amazing devices in our pockets one of these days.

Common Technical Characteristics of IMT-2000 3G Systems

To provide the features promised by IMT-2000, terminals and networks have some technical characteristics in common:

- ✓ High-speed data transfers
- ✓ Packetized data
- ✓ Internet protocol (IP) support

Each of these characteristics is covered in more detail in the following sections.

HIGH-SPEED DATA TRANSFERS

Most people think of high speed when they think of 3G. That certainly makes sense. The 2G systems in wide use today move data at around 10 Kbps. Speeds like that make the old 28.8 Kbps modem look pretty good.

The new 2.5G systems have improved speeds, ranging from 28.8 Kbps to as much as 300 Kbps (although a more realistic top range for 2.5G systems is closer to 144 Kbps). Keep in mind that 144 Kbps is a faster connection than most people have on their home PCs. A speed like that can support some multimedia and other demanding applications.

3G systems will be faster still. The bottom expected data rate for 3G is 384 Kbps, with the top end around 2 Mbps (2 *million* bits per second). At those rates, all sorts of exciting applications are possible, and tasks such as surfing the Web or downloading a large spreadsheet (even an MP3 file) to your data terminal become practical and painless.

While 3G systems will eventually deliver 2 Mbps (2.048 Mbps, to be exact), you still won't be able to get a 2 Mbps signal in your car while traveling down the autobahn at 150 kilometers per hour (kph). The velocity of the data terminal relative to the base station has a major impact on the data rate that can be achieved. IMT-2000 systems will be able to deliver 2 Mbps in low-mobility applications (when the data terminal is moving at less than 10 kph). For limited-mobility applications (when the relative velocity is less than 150 kph), the systems are required to work at 384 Kbps. For full-mobility applications (while the data terminal is in a moving vehicle traveling at less than 250 kph), the systems are required to deliver at least 144 Kbps, which is still an impressive feat.

PACKETIZED DATA

When a 3G system transmits data, it doesn't just blast it across the air interface in a continuous stream. Instead, the data is broken up into *packets*, discrete chunks of data with routing information and perhaps some error detection and correction bits thrown in.. Once the appropriate data terminal receives the packets, the data gets reassembled into its original form. Networks that use packetized data are called *packet-switched* networks. In contrast, earlier technologies are *circuit-switched*, meaning that each 2G wireless handset completely ties up one of the channels in a wireless cell for as long as the handset is connected to the network, whether or not the handset is actually transmitting on the channel.

Packetized data offers several benefits that make it ideal for use in 3G systems. First, because 3G systems are packet-switched, their data terminals can stay connected to the system at all times, sharing channels and only using any of the system's limited channel capacity when they are actively transmitting data. 3G data terminals are always-on devices. Thus, the terminal doesn't need to dial into the system and establish a circuit every time it wants to send some data. Also, it can receive data at any time, instead of only when the user remembers to connect the device to the system.

Another benefit of packetized data is that it simplifies error correction. If a packet of data is irretrievably corrupted, a terminal can easily request that the originating terminal retransmit the packet. This is impossible in a system where data travels as a continuous stream.

The switch to packetized data has two important implications that often go unstated. The first is that data becomes as important (or more important) than voice for 3G systems. Technologies like GSM work perfectly well for voice, without using data packets. Such technologies even allow for low-speed data transfers without all the extra work involved in packetizing data and extracting it from packets at the receiving terminal. Packets only become important when there is a lot of data to move on the system.

The other implication of 3G systems using packets is that all data on the system is now digital. Analog data (for example, voice and music) cannot be packetized without first converting it to a digital form. Packets are less efficient for analog data, so there needs to be a significant amount of digital data traveling on the network to make packetization worthwhile.

Supporting Wireless in the Next Generation of IP

Currently, the Internet protocol used on the Internet is *IP version 4*, or *IPv4*. IPv4 supports 32-bit addresses. Each address corresponds to a device or a collection of devices connected to the Internet. With 32 bits for addresses, IPv4 can uniquely specify over 4 billion addresses. While that sounds like a large number, the truth is, we are running out of addresses. Part of the problem is the expected massive proliferation of wireless devices that are connected to the Internet.

To correct this problem, in 1994 the Internet Engineering Task Force approved *IP version 6* (also called *IPv6*, or *IPng*, for *IP next generation*). Among its impressive qualities, IPv6 supports 128-bit addresses. This is an *extremely large* number, something like *80 quadrillion* times the number of addresses that IPv4 can support.

One publication, trying to put this number in perspective, suggested that approving IPv6 is like giving an IP address to every grain of sand on the planet. I think they might have underestimated, but the key is that IPv6 provides more IP addresses than we are ever going to need.

IPv6 has features that make it ideal for mobile wireless applications. IPv6 hardware can automatically detect changes to the local network and reconfigure itself to handle them. In particular, IPv6 data terminals will be able to roam across wireless networks (and parts of networks) much more efficiently than IPv4 terminals can.

Quality of service (*QoS*) is another issue that IPv6 will address. Currently, all data packets traveling across the Internet are treated as if they are of the same importance. Thus, a packet from an instant message about a dog's pedicure is treated as if it were just as important as a packet from a CIA report on terrorist movements overseas. The idea behind QoS is that the CIA should be able to specify a higher quality of service to its reports than I would assign to an instant message.

For a less dramatic example, think of a videoconference being broadcast to all of a company's satellite facilities using the Internet. Given the current system, where all packets have the same priority, odds are good that the satellite offices will experience jumps in the video or glitches in the audio, as packets containing parts of the conference get delayed or take different routes through the Net or need to be retransmitted due to an error. If the company could specify an appropriate QoS level for all packets associated with that videoconference, the quality of the presentation at the satellite offices would be much higher, if not flawless.

In short, the higher a packet's QoS, the higher its priority in the network. Higher priority packets will have a higher probability of getting to their destinations — and getting there faster. IPv6 will make that possible.

Varying QoS levels are also appealing to the network providers, as they can then charge more for higher QoS levels.

Expect IPv6 to roll out over the next several years, with the needs of wireless devices largely driving this rollout.

INTERNET PROTOCOL (IP) SUPPORT

The Internet is the most popular data network on earth. As such, it is the network that 3G systems should be able to connect with. The Internet protocol defines the format of data packets that travel on the Internet, and specifies the addressing scheme used. 3G networks support IP, so 3G networks can easily and efficiently exchange data packets with the Internet.

All the 3G technologies approved by the ITU for IMT-2000 support the Internet protocol.

IMT-2000 3G Technologies

As mentioned earlier in the chapter, three 3G technologies — WCDMA, UMT-136, and cdma2000 — are currently approved as part of IMT-2000. These ground-based technologies are sometimes collectively known by the acronym *UTRA*, which stands for *UMTS Terrestrial Radio Access*.

Wideband Code Division Multiple Access

WCDMA, also written *W-CDMA*, stands for *Wideband Code Division Multiple Access*. WCDMA was selected by the European Telecommunication Institute (ETSI) as the 3G technology that GSM would evolve into. As of now, it appears that most wireless companies (including NTT DoCoMo, a Japanese company that runs the hugely popular I-Mode service) are moving toward WCDMA as their 3G standard.

Table 10-1 shows the key characteristics of WCDMA technology.

TABLE 10-1 WCDMA Technology at a Glance

Characteristic	Value
Frequency Bands	1920–1980 MHz and 2110–2170 MHz
Necessary Bandwidth	5 MHz
Spectrum Division	CDMA
Responsible Organization	3GPP

Figure 10-1 shows where WCDMA fits into the overall evolution of 3G technologies. Note that, like GSM systems, some TDMA-based systems are also migrating to WCDMA.

Thanks primarily to its wide channels (5 MHz each), WCDMA offers several advantages over earlier commercial CDMA technologies, including the following:

- ✓ **Speed:** WCDMA can support the IMT-2000 requirement of data rates up to 2 Mbps in situations where the data terminal is moving slowly.
- ✓ **Higher capacity:** WCDMA can support twice as many users in the same bandwidth as earlier CDMA systems, and seven times as many users as standard GSM systems.

✓ **Greater immunity to signal fading:** Because the user signal is spread across more frequencies, fading at any given frequency has less of an impact on the overall signal.

In addition, WCDMA is designed to allow handoffs to and from GSM systems. This will be particularly important early in the deployment of WCDMA, when WCDMA coverage is still spotty and GSM is still widespread.

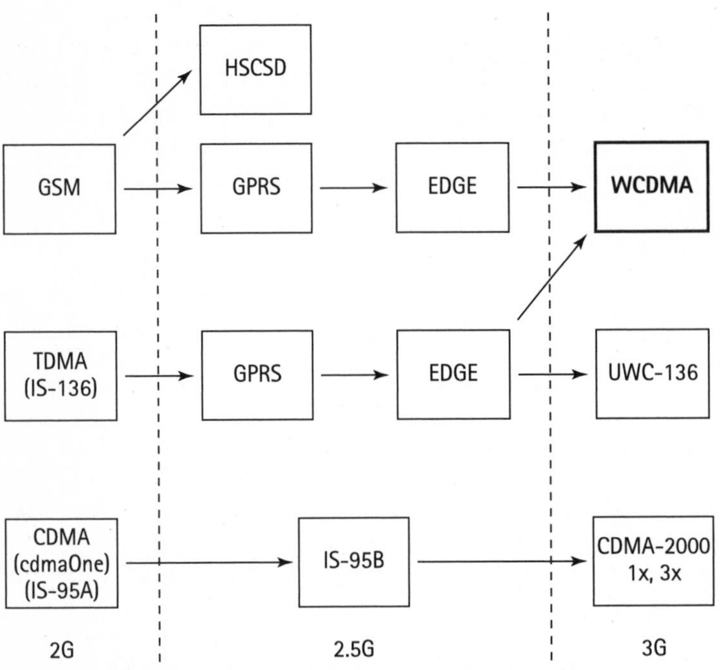

Figure 10-1: WCDMA is the 3G technology that GSM-based, and perhaps other, systems will migrate to.

WHO IS RESPONSIBLE FOR WCDMA?

The organization responsible for the creation, development, and maintenance of globally applicable technical specifications and technical reports governing WCDMA is the 3G Partnership Project (3GPP). 3GPP charter was actually meant to create the preceding kinds of documents for a "...3rd Generation Mobile System based on evolved GSM core networks and the radio access technologies they support...." In addition, 3GPP develops similar documents for versions of GSM that include GPRS and EDGE.

The 3G Partnership Project (3GPP) provides huge amounts of highly technical documentation on GPRS, EDGE, and WCDMA. Their Web site can be found at `www.3gpp.org`.

In short, the 3G Partnership Project is responsible for all the specs and technical reports that define and control the evolution of GSM into a 3G system. 3GPP members include interested organizations from around the world, such as the following:

- ✓ Association of Radio Industries and Businesses (ARIB) — Japan
- ✓ Chinese Wireless Telecommunication Standard group (CWTS) — China
- ✓ European Telecommunications Standards Institute (ETSI Secretariat) — European Community
- ✓ Committee T1 — North America
- ✓ Telecommunications Technology Association (TTA) — Korea
- ✓ Telecommunication Technology Committee (TTC) — Japan

HOW WCDMA WORKS

The basic principles of WCDMA are the same as those of current CDMA systems, and are covered in Chapter 8 under the heading "How CDMA Works."

Testing, Testing, 1-2-3

NTT DoCoMo began the first public test of WCDMA technology on May 30, 2001. This test program, called *FOMA* (*Freedom of Mobile Multimedia Access*), included 4,500 volunteers in and around Tokyo who received new handheld terminals capable of running at 384 Kbps. The testers reported incredibly high performance and better voice quality than on the systems they were used to.

These tests also turned up some bugs and problems (instability of the signal and low battery life) that needed fixing. No one should be surprised by this. As anyone who works in a high-tech organization knows, rolling out complex, cutting-edge technology to thousands of real users always turns up problems.

Commercial launch of the new service is estimated to occur around the end of 2001.

Another version of WCDMA uses TDD to separate signals originating at the base station from those originating at the data terminals. I do not know of any systems that use this variety of WCDMA today, but you should know it exists.

This section describes the major differences in the way WCDMA works, and takes a closer look at how a Direct Sequence Code Division Multiple Access (DS-CDMA) system such as WCDMA can distinguish one signal from the other.

One significant difference in the implementation of WCDMA is that the available bandwidth for 3G is large enough to allow multiple 5 MHz channels of WCDMA communication, each supporting multiple simultaneous users. Earlier generations of CDMA used a single channel.

Figure 10-2 displays a two-user example of this type of multi-user support. It portrays a tiny portion (3 bits) of a transmission from each user's data terminal to the base station.

At the top of Figure 10-2, you can see two three-bit chunks of data that are destined for the base station, and below, the segment of the pseudorandom code that gets applied to each user's data. At the bottom of the figure, you can see the resulting coded signal that will be transmitted for each user.

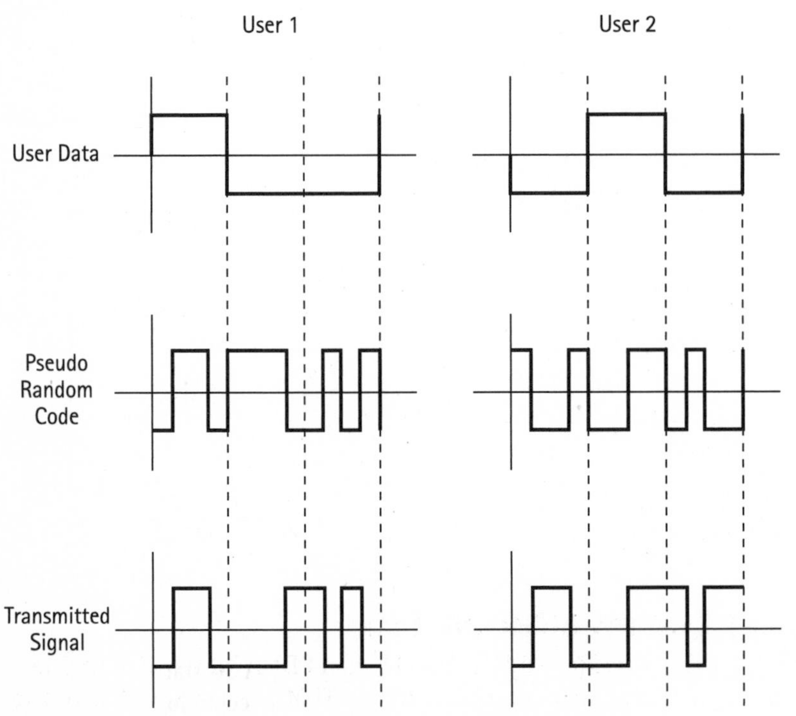

Figure 10-2: This figure portrays the encoding of two tiny chunks of data that are destined for two different data terminals in a WCDMA network.

Since all the data terminals are transmitting on the same channel, their signals get combined into an overall signal that is received by the base station. Figure 10-3 shows what happens when the base station receives the combined signal. By applying the pseudorandom code for each data terminal to the overall signal, the base station can reconstruct the original data from each data terminal.

This system works fine when the signals are of roughly the same power, and when there is little correlation between the pseudorandom codes. The problem of closer or more powerful stations overwhelming more distant or less powerful stations is known as the *near-far problem*. This is handled through the CDMA power control ability discussed in Chapter 8.

The *correlation problem*, on the other hand, is addressed primarily by the proper design of the pseudorandom code sequence. The correlation of the pseudorandom codes when actually used by multiple data terminals in the real world is also affected by the different time delays caused by the fact that the terminals are of varying distance from the receiving base station. While this means that the design of the pseudorandom codes cannot completely eliminate correlation effects in practice, the system can be (and is) designed to deal with this problem successfully.

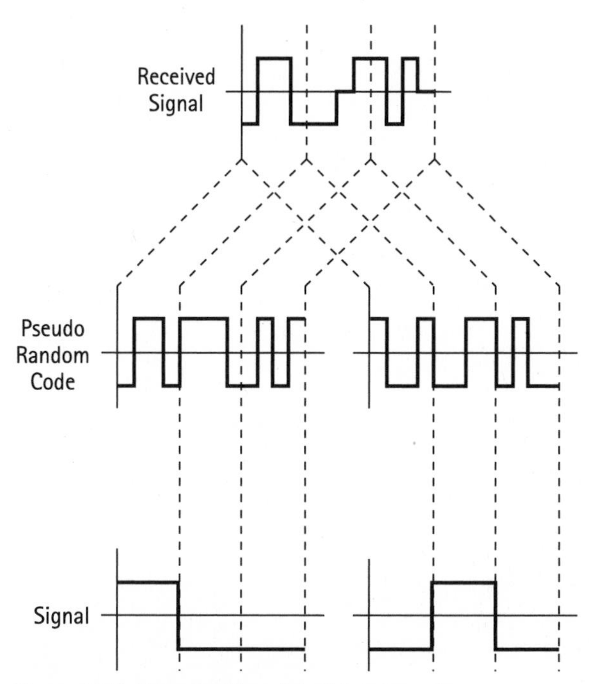

Figure 10-3: Extracting the original data from the received signal in a WCDMA system.

Universal Wireless Communication-136

UWC-136 (Universal Wireless Communication-136) is the only non-CDMA 3G standard approved by the ITU (at least at the time of this writing). UWC-136 is a TDMA technology designed to provide an evolutionary path to 3G for AMPS and AMPS-compatible (TIA/EIA-136) 2G systems, as well as for GSM. As such, UWC-136 offers some clear benefits for network operators. It specifically aims to provide a low-cost, incremental, evolutionary path for existing TDMA systems. UWC-136 can be deployed across a wide range of frequencies (500 MHz to 2.5 GHz, or even wider), so operators can conceivably offer 3G services in frequency bands that they now use for older technologies, thereby avoiding the bandwidth battles that are waging over the 3G portion of the spectrum. And because UWC-136 has no CDMA aspects, it can provide 3G speeds using channels as narrow as 1 MHz.

Figure 10-4 shows the path that operators with existing TDMA networks can take to reach 3G using UWC-136.

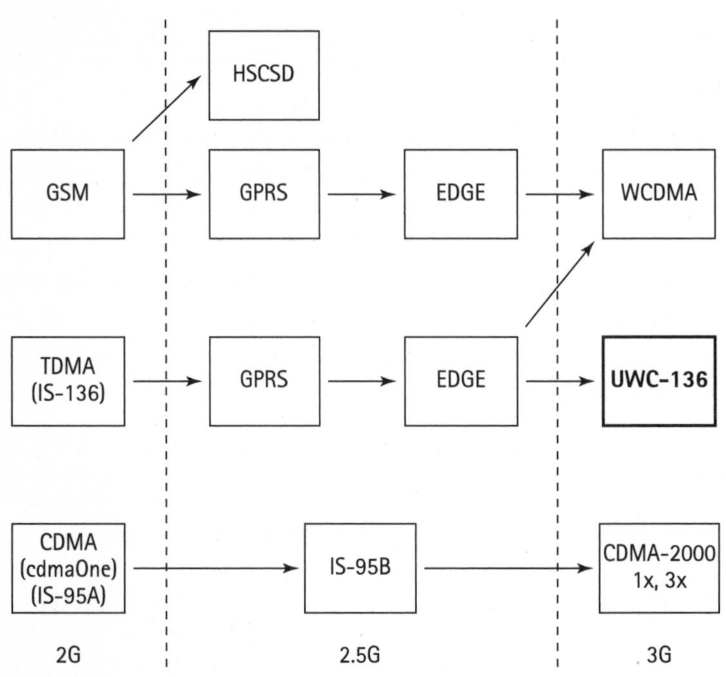

Figure 10-4: This figure shows where UWC-136 sits in the overall evolution of wireless to 3G technology.

The leading proponent of UWC-136 is the UWCC (Universal Wireless Communications Consortium). Their Web site is at `www.uwcc.org`.

Table 10-2 shows the key characteristics of UWC-136 technology.

Table 10-2 **UWC-136 Technology at a Glance**

Characteristic	Value
Frequency Bands	Any from 500 MHz and 2500 MHz
Necessary Bandwidth	1 MHz
Spectrum Division	TDMA
Responsible Organization	UWCC

Despite its real benefits, the fate of UWC-136 is unclear. The evolutionary path that leads from AMPS to UWC-136 includes GPRS and EDGE, two 2.5G technologies. But many network operators appear to be implementing GPRS and Edge as a direct path to WCDMA, instead of UWC-136. If you are interested in UWC-136 technology and its more incremental and economical approach to 3G, make sure to visit the UWCC Web site and pay attention to the relative adoption rates of WCDMA and UWC-136.

HOW UWC-136 WORKS

UWC-136 is fascinating in that it takes a very pragmatic approach to getting the job done. It includes the existing 1G and 2G TDMA technologies as modes within the overall specification. And it adapts GPRS and EDGE technologies to TDMA systems, thereby gaining their 2.5G performance benefits. At the top end, UWC-136 will make 3G speeds and capabilities available within as little as 1 MHz of bandwidth.

Keep in mind that UWC-136 isn't a single technology. It is a family of technologies, consisting of the standard 2G IS-136 technology, an enhanced version of the existing 136 technology, called 136+, and two higher bandwidth versions (136 HS and 136 HS Indoor) to handle the higher speed requirements of IMT-2000. Since IS-136 is addressed in Chapter 8, this section addresses 136+ and both types of 136 HS.

HOW 136+ TECHNOLOGY WORKS

136+ technology provides the following benefits:

- ✓ Packet-switched data at up to 43.2 Kbps
- ✓ Multiple modulation schemes to provide the best performance given the radio environment
- ✓ Use of the same 30 KHz channel bandwidth as IS-136

136+ achieves these benefits through its enhanced modulation schemes and variable bit rates. For environments where congestion and interference are likely, 136+ uses a coherent *QPSK modulation* scheme and the IS-641 vocoder. This combination produces a signal that is well suited for a high-interference environment by increasing the robustness (error resistance) of the signal.

Understanding PSK Modulation Techniques

While amplitude modulation works by varying the amplitude (level) of the carrier wave, and frequency modulation works by varying the frequency of the carrier wave, *Phase Shift Keying (PSK)* modulation techniques work by altering the phase of the carrier wave. PSK, in its most basic form, entails modulating the carrier wave by multiplying it by the digital signal you wish to encode. In this approach, each bit on the binary signal is represented by a 1 or a –1, instead of a one or a zero.

The result of modulating the carrier wave by the binary signal is an output wave that has its sine wave shifted by 180 degrees wherever a binary 0 appears in the original digital signal. The following figure illustrates this with a simple input signal.

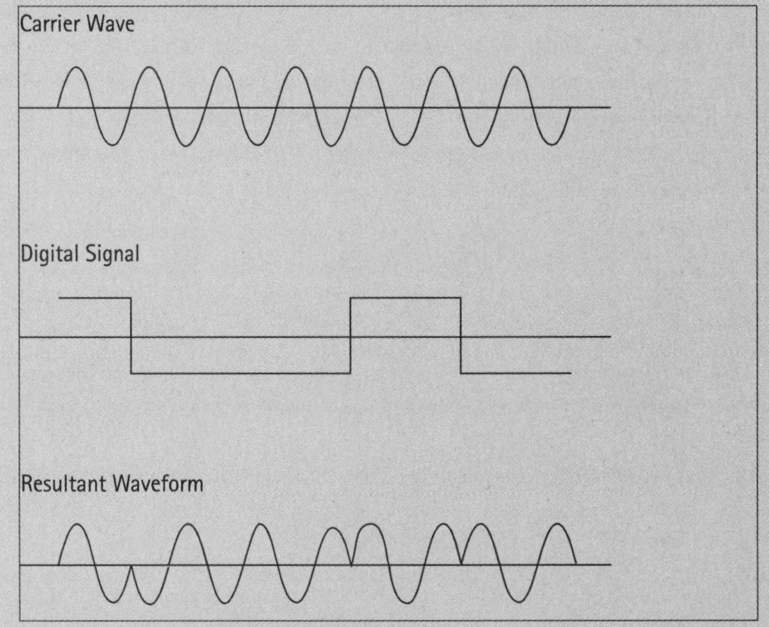

PKS modulation via combined waveforms

The resulting waveform is actually the sum of two waveforms, one representing the carrier wave, the other representing the binary signal. By filtering out the higher frequency carrier wave, a receiver can reconstruct the original binary signal.

With PSK, you can encode one bit of data in each phase change. However, with more sophisticated equipment, you can encode, and subsequently decode, phase changes of less than 180 degrees. 136+ uses *Quadrature Shift Phase Keying (QPSK)*, which can apply four phase changes to the carrier wave, and 8-PSK, which can apply eight phase changes to the carrier wave.

> Of course, the smaller the phase change, the harder it is to detect, particularly in the real world — where interference exists, signals drift slightly, and nothing functions as perfectly as it does on paper. Thus, 136+ uses QPSK in congested, interference-prone areas, and the more efficient but less robust 8-PSK technique in areas where there is likely to be less interference.

For environments where congestion and interference are less likely, 136+ uses a coherent *8-PSK modulation* scheme with the US1 vocoder. This combination produces higher fidelity voice signals at the expense of a less robust signal.

HOW 136 HS TECHNOLOGY WORKS

In contrast to 136+, 136 HS technologies are data-only. They do not support voice communication. This makes sense, as the HS versions work at rates far higher than those needed for voice signals. The two versions of 136 HS have the characteristics, as shown in Table 10-3.

TABLE 10-3 **136 HS Outdoor and 136 HS Indoor at a Glance**

Characteristic	136 HS Outdoor	136 HS Indoor
Supported Data Rates	64 Kbps to 384 Kbps	384 Kbps to 2 Mbps
Necessary Bandwidth	200 KHz	1.6 MHz
Modulation Schemes	8-PSK, GMSK	BOQAM, QOQAM
Supported Environments	Pedestrian and vehicular at speeds in excess of 100 kph	Indoor office use only

136 HS Outdoor is designed to support data rates of up to 384 Kbps, even when the data terminal is moving at speeds up to 100 kph. This exceeds the requirements of IMT-2000.

cdma2000

cdma2000 is one of the two CDMA standards approved by the ITU for inclusion in IMT-2000 (the other being WCDMA). cdma2000 is the preferred 3G solution for networks currently based on cdmaOne, as the two are forward and backward compatible. The adoption of cdma2000 preserves the investments that network operators (primarily in North America and Korea) have made in cdmaOne systems.

Figure 10-5 shows where cdma2000 stands in the evolution of wireless systems to 3G.

Revisiting PSK Modulation Techniques

Part of the speed gain in 136 HS technologies is due to the modulation techniques they use. In addition to the 8-PSK scheme used by 136+, these technologies also use *Gaussian Minimum Shift Keying* (*GMSK*), *Binary Offset Quadrature Amplitude Modulation* (*BOQAM*), and *Quaternary Offset Quadrature Amplitude Modulation* (*QOQAM*).

GMSK applies a Gaussian filter to a *Minimum Shift Keying* (*MSK*) signal, thereby reducing the spectral density of the signal and allowing for more efficient use of the channel bandwidth. Minimum Shift Keying encodes binary data on the carrier wave by leaving the carrier wave unaffected for a binary one, and by increasing the frequency of the carrier wave by half the data rate for a binary zero. The frequency changes occur at the zero crossing points of the carrier wave, and there are no discontinuities in the phase of the carrier wave. The following figure should clarify this point.

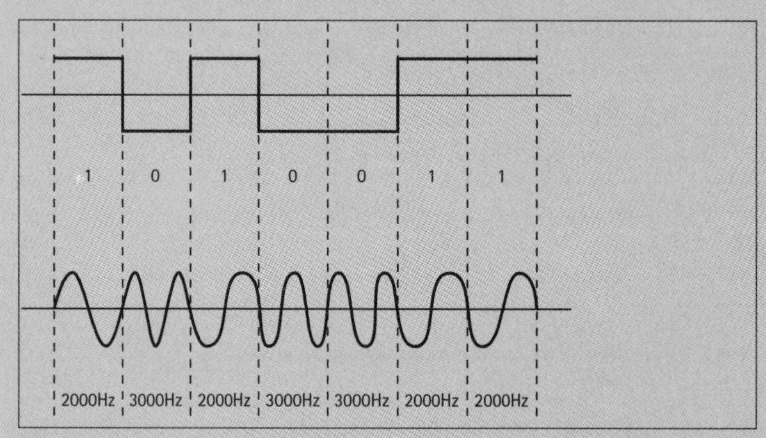

PKS modulation via *Gaussian Minimum Shift Keying*

QAM takes a very different approach to modulation. In QAM, the carrier wave is comprised of two sinusoidal sine carrier waves, each with the same frequency, but exactly 90 degrees out of phase. These waves can be modulated independently, then successfully demodulated at the receiver. The resulting system allows twice the data to move through the channel (since there are effectively two independent carrier waves within the channel) without any increase in error rates.

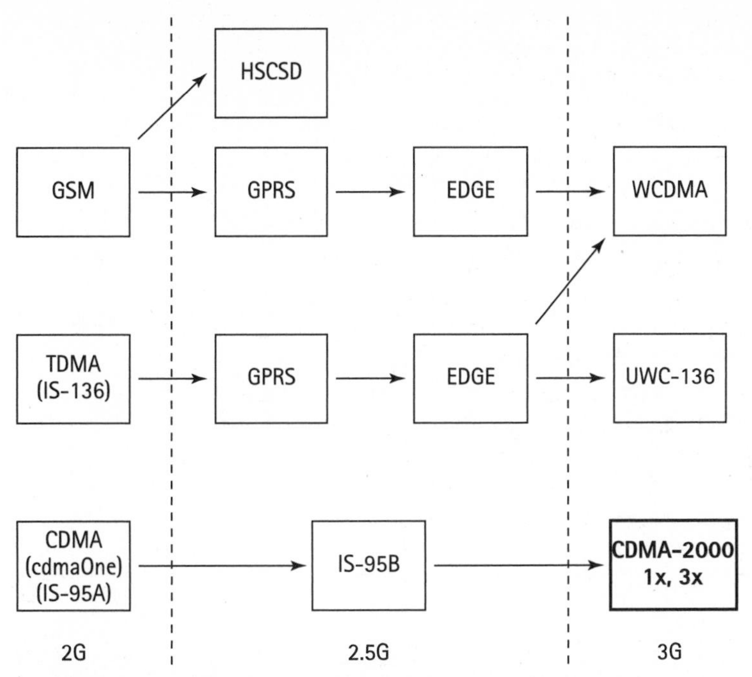

Figure 10-5: cdma2000 is the 3G technology that most cdmaOne systems are moving to.

cdma2000 is also known as *CDMA-Multi-Carrier*. This name refers to the fact that cdma2000 is designed to comply with ANSI document TIA/EIA-41, which specifies the IS-95 and IS-136 2G technologies.

The organization responsible for the overall cdma2000 standard is the 3G Partnership Project 2, also known as 3GPP2. Their Web site is found at `www.3gpp2.org`.

Table 10-4 shows the key characteristics of cdma2000 technology.

Table 10-4 cdma2000 Technology at a Glance

Characteristic	Value
Frequency Bands	1920–1980 MHz and 2110–2170 MHz
Necessary Bandwidth	1xRTT: 1.25 MHz
	3xRTT: 3.75 MHz
Maximum Data Rate	1xRTT: 14.4 Kbps on IS-95A networks; 114 Kbps on IS-95B networks; 307 Kbps on 1x and 3x networks.
	3xRTT: 14.4 Kbps on IS-95A networks; 114 Kbps on IS-95B networks; 307 Kbps on 1x networks; 2 Mbps on 3x networks.
Spectrum Division	CDMA
Responsible Organization	3GPP2

cdma2000 deployment began late in the year 2000, with 1xRTT commercially available in Seoul through two operators. Numerous additional deployments were originally scheduled for the second half of 2001, but, given the economic climate, it is unclear whether they will occur on schedule.

THE PHASES OF CDMA2000

cdma2000 might best be considered a series of closely-related and ever more capable standards and specifications. The exact details are somewhat confusing — even the exact names of the phases vary, depending on which source you consult. Table 10-5 should help bring the big picture into focus.

Table 10-5 The cdma2000 Family of Standards

Name	Part of IMT-2000?	Description
1xMC, 1xRTT, MC1X, cdma2000 1X, 3G1X	Yes	First phase of cdma2000 implementation. Data rates of over 144 Kbps, along with twice the voice capacity. Can operate on cdmaOne networks at rates consistent with those networks. Can also run on later networks.
3xRTT, cdma2000 3X, 3xMC, MC3X	Yes	Second phase of cdma2000 implementation, as approved by the ITU for IMT-2000. Uses 5 MHz of radio spectrum and technical improvements to deliver data at 2.4 Mbps. Can operate on cdmaOne networks at rates consistent with those networks. Can also run on later networks.

Name	Part of IMT-2000?	Description
cdma2000 1X EV-DO, 1X-EV Phase 1	No	An enhancement to 1xMC that splits voice and data onto separate channels, allowing data rates of 2.4 Mbps. Can operate on cdmaOne networks at rates consistent with those networks. Can also run on later networks.
cdma2000 1XEV-DV, 1X-EV Phase 2, EV-DV	No	An enhancement to 1xEV-DV that would provide data rates of 3 to 5 Mbps.

The first phase is 1xRTT, also known by the myriad names listed in Table 10-5. It provides data rates of over 144 Kbps, along with twice the voice capacity of networks running cdmaOne. At the same time, 1xRTT and cdmaOne data terminals are compatible with each other's networks. 1xRTT terminals will work on cdmaOne networks, although they are limited to working at the speed of the network. cdmaOne terminals will work on 1xRTT networks as well. This ability to hand off between the networks will be very important early in the cdma2000 adoption cycle, when cdma2000 coverage is incomplete, and cdmaOne coverage is still more widespread. The 3G network that Verizon Wireless deployed in the New York City area in August 2001 was based on 1xRTT.

The second phase is 3xRTT, also known as 3xMC, cdma2000 3x, or MC3X. This phase expands on 1xRTT by adding support for varied channel sizes, data rates of up to 2 Mbps, and advanced multimedia capabilities.

The cdma2000 1XEV-DO and 1XEV-DV approaches are not part of the IMT-2000 plan. Aside from technology demonstrations, these variants on cdma2000 exist primarily as specifications and proposals.

HOW CDMA2000 WORKS

Since it was designed in part to protect network operator's investments in existing cdmaOne networks, cdma2000 needs to support IS-95 data terminals within a cdma2000 network. To achieve this, the length of each data frame is changed from that of WCDMA, and the *chip rate* is modified to be three times that of IS-95. With these changes, it becomes practical for cdma2000 to support IS-95 carrier signals as one mode of operation beyond the main mode — hence the name *CDMA-Multi-Carrier*.

cdma2000 3xRTT appears to be very much a work in progress. Therefore, this section concentrates on 1xRTT, which has already been deployed, and is reasonably stable.

See the heading "How CDMA Works" in Chapter 8 for the basics on this technology. While many details of cdma2000 differ from those of earlier CDMA systems or WCDMA, the basics remain the same.

In the 1xRTT mode, cdma2000 is an integrated voice and data service. Instead of separating data and voice and handling them independently, 1xRTT treats voice exactly the same as data. The "1x" in the name results from the fact that 1xRTT uses the same 1.25 MHz channel bandwidth that cdmaOne systems use.

Because it uses the same size channels as cdmaOne, cdma2000 can be integrated into existing wireless systems, replacing existing channels as required. 1xRTT includes more sophisticated control of the data terminal power levels by the base station, additional modulation techniques for the *reverse channel,* and improved data coding methods. The increased efficiency of the design allows for twice the voice capacity on a given channel, along with data rates of up to 144 Kbps. The improved terminal power control means even better battery life for devices.

To obtain more technical details about cdma2000, you'll need to attend formal training on the subject. Examples of courses include Qualcomm's CDMA University (`www.cdma.com/cda/technology/training/`) and the UCSD Extension online course in cdma2000 (search for "cdma2000" at the Extension's site: `www.extension.ucsd.edu/`).

Another source of detailed technical information on cdma2000 is the archive of standards documents at the TIA Web site (`www.tiaonline.org`). A search on cdma2000 at this site will bring up a complete list of the relevant standards documents, which you can then purchase. At this writing, you could purchase all the relevant documents for around $1,000.

Related Technologies

As you might expect in any field where rapid innovation is the rule and huge amounts of money can be made, the 3G standards approved by the ITU for IMT-2000 are not the only technologies vying for a piece of the future in wireless communication.

The following sections discuss two technologies that may have a big impact on the future of wireless devices. The first, TD-SCDMA, is an alternative 3G technology that was promoted by Chinese manufacturers and others as a possible IMT-2000 standard. While it did not make the original list, TD-SCDMA is still being developed, and may yet become a major player in 3G.

The other technology, smart antennas, isn't a direct competitor to any of the 3G air interface standards. Instead, it is a technology that could complement any or all of them, providing additional speed and capacity to their networks.

Time Division Synchronous Code Division Multiple Access

TD-SCDMA, or *Time Division Synchronous Code Division Multiple Access*, is a 3G technology being developed by China, in association with the German company Siemens AG. While not approved by the ITU as part of IMT-2000, TD-SCDMA is being designed to many of the IMT-2000 standards to allow for easier interoperability with WCDMA, in particular. This, combined with the fact that China is already one of the largest mobile phone markets in the world, makes TD-SCDMA a technology worth looking at. TD-SCDMA is sometimes also called *TDD Low Chip Rate*.

The place to go for information on TD-SCDMA is, not surprisingly, the TD-SCDMA Forum Web site. The site has both Chinese and English versions, so I've included the link to the English version here: `www.tdscdma-forum.org/english/index.html`.

Table 10-6 shows the key characteristics of TD-SCDMA technology.

TABLE 10-6 **TD-SCDMA Technology at a Glance**

Characteristic	Value
Frequency Bands	Various
Necessary Bandwidth	5 MHz, divided into three 1.25 MHz or 1.6 MHz channels
Maximum Data Rate	384 Kbps
Spectrum Division	TDMA and CDMA
Responsible Organization	Ministry of Information Industry (China)

TD-SCDMA ADVANTAGES AND DISADVANTAGES

The backers of TD-SCDMA claim many benefits for the technology: increased spectral efficiency compared to WCDMA, the ability to work with paired or unpaired channels, the ability to support asymmetrical uplink and downlink channels (which allows for higher bandwidth in one direction, usually down to the data terminal), and an integrated location service built on smart antenna technology.

In addition, TD-SCDMA can work with existing GSM networks by means of hard handoffs between the two. This is a major advantage when you consider that GSM is the dominant wireless technology in China today.

For detailed information about smart antennas, refer to the section at the end of this chapter titled "Smart Antenna Technology.".

On the other hand, TD-SCDMA has some disadvantages. The maximum data rate planned for the technology is 384 Kbps. It will not (at least as currently described) function in fast-moving vehicles, and the range of base stations will probably be much shorter than for pure CDMA technologies.

TD-SCDMA is fortunate to have the backing of the Chinese government. Current indications are that the government wants to deploy a Chinese 3G technology, and that Chinese companies will comply with this desire. So TD-SCDMA is almost certain to be deployed in one of the biggest markets on earth, giving it a strong position as one of the 3G standards of the future.

HOW TD-SCDMA WORKS

You can think of TD-SCDMA as a TDMA system where multiple users share each time slot using CDMA. In normal operation, TD-SCDMA is designed to have seven time slots per frame, with up to sixteen users per time slot.

The receivers use a technology known as *joint detection* to discriminate between the multiple CDMA users in each frame. Joint detection employs a multidimensional matrix calculation to eliminate the interference effects of the CDMA signals on each other, improving the efficiency of the system. Due to exponentially-increasing computation requirements, joint detection is only feasible for CDMA systems where there are relatively few CDMA signals to detect. TD-SCDMA allows only sixteen simultaneous CDMA signals, making the use of joint detection possible.

Joint detection allows TD-SCDMA networks to employ a frequency reuse factor of three, instead of the more common seven. This allows the network to support more users per base station, reducing overall network costs.

See the heading "Frequency Reuse" in Chapter 7 for more information.

TD-SCDMA is designed to be very flexible in the way that it handles signals. For voice and other low data rate applications, the system functions in its normal mode. For high data rate applications such as multimedia, TD-SCDMA is designed to go into a full TDMA mode. This enables higher data rates, at the price of a reduced number of simultaneous users.

TD-SCDMA employs a smart antenna system to further reduce interference, improve signal strength, and potentially aid in determining the location of the users. In this application, the base station uses the signal strength to determine the distance to the data terminal. The next section provides general information about how smart antennas work and how they determine the direction (bearing) of the data terminal. With a bearing, distance from the base station, along with the known location of the base station, the TD-SCDMA network can compute the location of the data terminal.

To make this computing and transmitting work smoothly, and to simplify upgrades to the system, TD-SCDMA radios will be *software defined radios* (*SDRs*). Current data terminals use custom microelectronics to handle most of their functions — in particular, the radio frequency functions such as modulation and demodulation. In an SDR, digital signal processors, high-speed microprocessors and software replace the custom microelectronics.

A SDR will enable the network operator or terminal manufacturer to reprogram the device to take advantage of new technologies or improvements to existing technologies. Software radios are expected to become practical over the next few years.

For the latest information on software defined radios, visit the Software Defined Radio Forum Web site at `www.sdrforum.org/index.html`.

Smart Antenna Technology

While most efforts at improving wireless devices are directed at improving protocols, transmitters, and receivers, dramatic performance gains are possible in another area: antennas. *Smart antenna systems*, which already deliver data at 64 Kbps to some lucky mobile users in Taiwan, offer another path to high-performance wireless devices.

If you want general information on smart antenna technology, start with the International Engineering Consortium (IEC) Smart Antenna Systems Tutorial Web site. This tutorial, which was created with information from ArrayComm, Inc., covers the major aspects of smart antenna technology. It can be found at `www.iec.org/tutorials/smart_ant/index.html`, and the ArrayComm Web site at `www.arraycomm.com`.

HOW SMART ANTENNA TECHNOLOGY WORKS

To understand — and appreciate — smart antenna technology, it is useful to understand the basic workings of traditional antennas

TRADITIONAL ANTENNAS Most wireless devices are equipped with what is known as an *omnidirectional dipole antenna*. Sometimes, the antenna is inside the casing of the device; other times, it protrudes from the device. But in either case, the antenna transmits and receives equally well in all directions perpendicular to its long axis. See Figure 10-6 for a side and top view of an omnidirectional dipole antenna's coverage pattern.

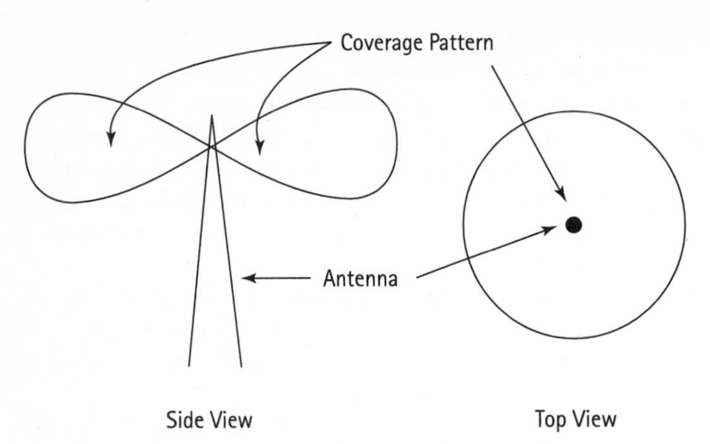

Figure 10-6: The coverage pattern for an omnidirectional dipole antenna

An omnidirectional antenna, while simple to implement and use, has some drawbacks. Because it receives signals from any direction with equal efficiency, interference is a real issue — even when the problem is mitigated by the design of cellular systems, with their varied frequency reuse schemes.

Transmitting using an omnidirectional antenna is also a serious problem. To communicate with the intended receiver, the antenna must broadcast with enough power to create the minimum required signal level at the receiver. Since the antenna is broadcasting in all directions equally, only a small percentage of the total broadcast power reaches the intended receiver. This means that most of the signal broadcast by the transmitter is wasted energy, or even interference that other nearby wireless devices must overcome.

SMART ANTENNAS A smart antenna is very different from an omnidirectional dipole. Smart antennas comprise an array of antenna elements, along with the electronics necessary to drive them in highly controlled ways. Such systems can alter their reception or transmission patterns under software control. In effect, by varying the way the signals associated with each element of the antenna system are processed, smart antennas can focus their "attention" in a particular direction. Just as your auditory system (ears plus brain) can concentrate on a particular conversation in a crowded room, effectively amplifying that conversation and ignoring others, so can smart antenna systems concentrate on a particular signal, amplifying it and rejecting interference. There are two main types of smart antenna systems under consideration for communicating with wireless devices. The switched-beam systems have a set of predefined beam patterns that they switch between, selecting the pattern that provides the greatest gain at any given time. Adaptive-beam systems can generate an infinitely variable number of beam patterns, dynamically modifying the shape of the beam to always get the best possible signal.

The physics and mathematics involved in the making of smart antennas are beyond the scope of this book. However, the concepts and techniques that smart antenna systems use are well proven. Besides the ArrayComm smart antenna system already moving data at 64 Kbps in Taiwan, the military has been using smart antennas for decades. Aircraft and ships are often equipped with phased-array radar systems, which are smart antennas used for detection of objects, rather than communication.

THE BENEFITS OF SMART ANTENNA SYSTEMS

Smart antenna systems are clearly more complicated and expensive than omnidirectional dipole antennas, but in return, they offer some real advantages for wireless systems. Because smart antenna systems can shape their coverage patterns to maximize the signal gain for any given signal, they can have larger coverage areas. At the same time, smart antenna systems can transmit with lower overall energy levels, since their beams are directed toward the intended receiver instead of broadcast omnidirectionally. Thus, the overall noise and interference they produce is diminished.

Another way to use the directional ability of smart antenna systems — implemented by ArrayComm for their system in Taiwan — is to establish *spatial channels*. By establishing these channels for each user, the system increases the effective carrying capacity of the wireless network and uses that increased capacity to provide higher data rates to users.

Smart antenna technology is an approach that can complement virtually any existing 2G, 2.5G, or 3G technology. As it matures, and as companies like ArrayComm get more field experience, smart antennas will likely appear in more and more wireless systems.

Summary

This chapter gives you a feel for the current state of 3G wireless technology. Although ITU is busy setting standards, and many companies and industry groups are working to create specific technologies that meet these standards, 3G is still very much in its infancy. The industry as a whole is pushing the limits of current technologies, trying to create the high-performance, world-spanning, Internet-in-your-pocket 3G world that we have come to expect. However, the full deployment of 3G wireless technologies is going to take several years. Along the way, we'll see more snags and problems, even downright failures. We're all going to have to be patient.

In this chapter, the following points were discussed:

- ✓ 3G is not a single technology; rather, it is a set of technologies that meet certain standards and are interoperable at a certain level.
- ✓ The international standard for 3G technologies is called IMT-2000, and it is administered by the International Telecommunications Union (ITU).
- ✓ Three technologies have now been approved as part of IMT-2000: WCDMA, UMT-136, and cdma2000.
- ✓ China is promoting its own 3G technology, TD-SCDMA, which may eventually be included in IMT-2000 or a subsequent standard.
- ✓ Smart antenna technology can be applied to 3G technologies to improve their efficiency, and may play an important part in future 3G solutions.

Chapter 11

Fixed Wireless and the "Last Mile"

In This Chapter

So far in this book, when wireless is mentioned, we also imply mobility. Wireless devices are mobile, and wireless solutions are geared primarily to accommodate mobility. However, wireless communication has another dimension as well: fixed wireless.

Fixed wireless is not a hot topic outside of the telco industry because of a lack of gadgetry involved. The term fixed wireless is used to describe multiple solutions to bring wireless communications to population centers, either remote ones or those that are hard to access. Fixed wireless deploys a wireless technology along a specified bandwidth between two or more cells, access points, or antenna.

- ✓ Fixed wireless network uses
- ✓ Types of fixed wireless networks
- ✓ Last mile solutions
- ✓ Fixed wireless network topologies
- ✓ Wireless redundant links

Fixed Wireless Uses

Fixed wireless technology is deployed in two broad categories, extending an existing network or developing a "last mile" solution.

Extending an Existing Network

Fixed wireless can extend a network across a distance where there is neither the motivation nor financial support to lay down a copper or fiber network. For example, if new commercial or residential areas spring up a distance away from a metropolitan area, they will likely require wireless services sooner, rather than later. You can quickly deploy a wireless network to accommodate this need.

If radio transmitters are already in place, wireless access points can be piggybacked alongside these transmitters, quickly extending bandwidth to the new neighborhood. Fixed wireless transceivers can also be deployed to beef up an overloaded telco network, as has been done in Eastern Europe.

You may wonder, "Why wouldn't it make sense just to lay fiber or copper lines?" Digging up ground to lay copper or fiber connections requires municipal approval on the part of many agencies. Environmental constraints and eco-damage potential can tie up wired network deployment for years. You can deploy a wireless network at least as a stopgap measure until a higher-speed, fixed wired network can be deployed.

The primary interest in putting up a fixed wireless network under the above circumstances is quick deployment. A wired network over a distance is always better, as data rates over copper — and especially fiber — will be much faster and can carry more data.

The specification used for these types of fixed wireless solutions is often 802.11b, relayed a long distance, using frequent access points and signal amplification. However, some solution providers such as DMC Stratex Networks offer very high speed (OC-3) microwave transmission, linked to fiber at one end, using advanced media converters. One solution, called *WTTC*, or *Wireless to the Curb*, provides a wireless microwave link to a station just outside the customer's plant or campus. These wireless/fiber solutions deliver both the high speed of fiber with the flexibility and non-invasiveness of wireless. Also available are wireless optical products for large campuses and urban areas from companies such as AirFiber and Lucent's WaveStar OpticAir. These are expensive, complex high-speed data-delivery systems utilizing lasers, amplifiers, and receivers over both air and fiber. Devices are placed in windows or rooftops, and the signal is amplified as needed and routed to optical fiber, if required. From there, data is transmitted longer distances than air would allow. These solutions are a departure from other point-to-point technology in that line of sight is not required.

A Last Mile Solution

Fixed wireless has a greater potential as a last mile or wireless local loop solution. *Last mile* refers to the link between an ISP's last big transceiver on the "edge of town," and the many customers' homes or offices. The best technology to deploy bandwidth right up to the "edge of town" would be a single fiber pair or copper underground. However, the technology required to get that bandwidth to the myriad citizens beyond that point would indeed be wireless (see Figure 11-1).

Before we get too carried away with this city analogy, we should note that last mile also refers to the need to link any large group of customers where it's not feasible to lay down wired lines. Last mile solutions are considered in the following environments:

- ✓ Universities and schools
- ✓ Large medical centers
- ✓ Townships and villages in developing nations
- ✓ Sacred lands
- ✓ Fragile ecosystem environments
- ✓ Any city not amenable to awakening to the sounds of armies of jackhammers digging up the streets

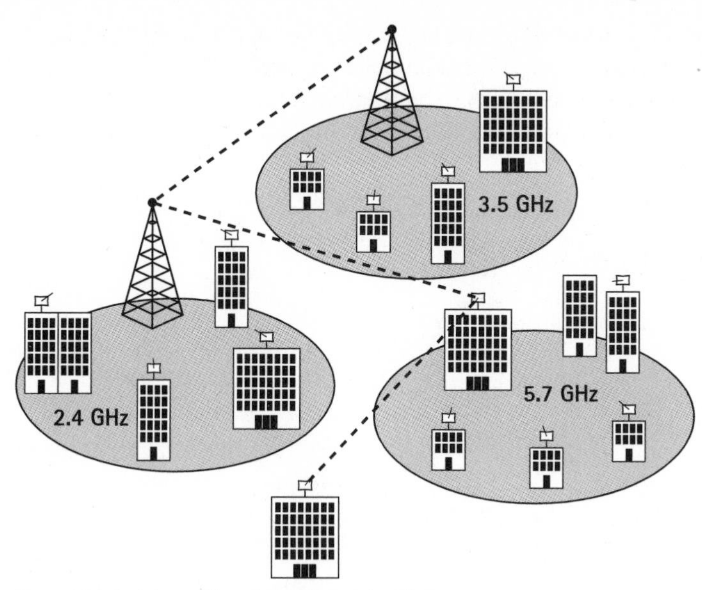

Figure 11-1: A last mile solution is required here.

This discussion is far from merely theoretical. It's quite conceivable that some of you reading this book could end up working for a telco in a small county someplace, suddenly commissioned to extend a network overnight to 50,000 new residents. Your expertise in this area will then be much appreciated. From the time that first modem's "data" light starts blinking, you'll be regarded a local hero.

Here are reasons why you'd hesitate to create wired network in such environments:

- ✓ A school or university, where students and faculty require wireless access to a network and data services, as well as the Internet. Universities often feature older, historical buildings. Drilling in walls and laying Category 5 wiring is likely to be not much of an option.
- ✓ A medical center, where wireless access to medical records, test results, library materials, a central network, and the Internet are all required. In such settings, service disruption in order to lay down cable is not allowable, so wireless deployment is preferable.
- ✓ A large plant, where inventory records, employee data, and other company information must be network-accessible. Of course, a Web portal can be provided as well. Again, most business models can't tolerate enough downtime to facilitate laying a wired network.
- ✓ A financial institution or corporate campus, which would require a variety of network services, including intranets, VPN, video conferencing, and other access requirements already named. In such situations, a wireless network is often used to supplement existing fixed wired Ethernet links.

- **A remote township, where inaccessibility and lack of physical infrastructure have hampered providing municipal services.** In such environments, wireless service can be a godsend, providing hitherto unheard of access to markets, medical supplies, and other provisions, whereas control of wired networks could often be in the hands of unfriendly national governments.
- **A large government agency that requires sharing of records, access to the Internet, and government files.** Again, the disruption caused by laying down a wired network is not workable.
- **Remote land bureaus or research outposts, where remote access to agency data is a must, but quarters are temporary.** In this case, the required presence is not permanent; hence, a wireless solution is preferable.

Note that when a wireless network is deployed over access points, the municipality or campuses themselves can have roaming access. If 802.11 technology is distributed throughout multiple buildings and grounds, wireless devices can enjoy untethered access. Throughout offices, classrooms, or hospital suites, users can access network services via mobile, wireless devices.

If users access the network with mobile devices, it's important that cells have frequent points where they overlap each other (similar to cellphone systems). This allows access to the network through more than one point, should everyone find themselves congregating and online in the same auditorium, for example.

Deploying Fixed Wireless Networks

There are two types of wireless local loop technologies: point-to-point and point-to-multipoint, sometimes called just multipoint.

Point-to-Point

Point-to-point is usually used to connect two large campuses. A group of corporate buildings within sight of each other on either side of an international border is the clearest example. This is called a dedicated link. Users can share a VPN connection, a network, an intranet, and a Web portal. Point-to-point wireless networks are often microwave, and, depending on the distance between campuses, require intervening access points. They are not cheap and usually are privately financed by large firms. Through clever deployment of access points and amplification, point-to-point wireless networks can extend up to 30 kilometers.

Point-to-Multipoint

Point-to-multipoint wireless networks are more common. One station or antenna provides bandwidth to many customer points.

When you use point-to-multipoint, the total bandwidth is shared between all users in a cell. However, service may not be uniform, because data is available asymmetrically.

Users closer to an available access point will have better throughput rates than those a bit farther afield. Also, if data use congregates in one part of the system (if everyone is in the dorms watching a Radiohead Webcast), that point of heavy use may not be able to draw the required bandwidth. This can be true even if the total bandwidth available to the network is sufficient. However, if access points are arranged in such a way so that high-traffic segments cannot grab more bandwidth from other points, bottlenecks can occur. To prevent this, you would have to deploy access points in such a way that stressed segments can "see" three or four other points, and use their bandwidth, if necessary. In this sort of deployment, you may have to plan ahead and predict areas where more bandwidth may be needed.

Fixed Wireless Technologies

There are several specifications, or technologies, used to transmit data on fixed wireless systems.

IEEE 802.11

The most common network technology on a local loop is a variant of 802.11 DSSS, deployed at the 2.4 GHz frequency. This technology provides a theoretical data exchange rate of 11 Mbps, but as distance between cells increases, the actual speed may be less. Use of this spectral range is desirable because it is unlicensed, and users would pay no reoccurring fees, as happens when operators are allowed to sell spectrum to customers. Vendors, however, can make some money and increase the actual data throughput for fixed wireless by offering proprietary solutions – solutions that tie the buyer into using vendor's equipment. Since a one-time investment in heavy merchandise is not apt to be repeated in the near future, customers of these solutions must choose carefully. Indeed, once installed, a big network operator will not replace stations in a point-to-point link, or even in a multipoint link, until the next generation of equipment becomes available. Hence, the fixed wireless network space has a healthy selection of vendors to choose from, each with their own bandwidth-enhancing spin. Companies such as InnoWave and HTE8 provide customized fixed wireless solutions using the right technology for your needs. These could involve, for example, proprietary spins on CDMA for wireless DSL, in tandem with a wired network.

For a discussion on another popular fixed wireless solution, see Chapter 10, specifically the "Smart Antenna Technology" section.

MMDS

The MMDS (Multichannel Multipoint Distribution System) frequency starts at 2150 MHz. In the mid-1970s, the FCC began setting aside this frequency for TV operators who would compete with cable. This frequency was unpopular as a television technology, and the FCC later sold portions of the MMDS spectrum to big carriers such as Sprint and WorldCom. Capable of distances up to 35 miles at its lower frequencies, users can usually hit a workable download transmission rate of 25 Mbps, which would certainly keep customers happy.

Fixed wireless Internet access service on MMDS is offered in several big residential markets, usually in areas where DSL is hard to come by. However, chronic carrier bankruptcies and technical hurdles have delayed once-anticipated huge MMDS-based service rollouts. Still, plans are being gradually revamped, and you'll probably see more MMDS-based service in the near future.

LMDS

Another very promising fixed wireless multipoint technology is LMDS, or Local Multipoint Distribution Services. Deployed at short range for densely populated areas, LMDS provides a peak rate of 155 Mbps. This technology can provide enough bandwidth for voice and Internet access, as well as online multimedia applications. As more carriers purchase licenses for available channels, you'll see more services deployed on LMDS.

Fixed Wireless Network Topology

Access point distribution formation — or network topology — affects network accessibility in a number of ways.

- ✓ **Star networks:** The most common type of LAN, and a very efficient fixed wireless network as well, star networks connect each user to a central site. Such arrangement helps prevent asymmetry. If one user is set farther away than others, or is behind an obstruction, another access point or signal amplification can be utilized to augment access.
- ✓ **Ring networks:** In a ring network, each access point or cell is connected to another in a ring formation around the area. This system promotes fault tolerance. In the event of an access point or cell failure, customers will still stay connected. That's because if one node fails, the user will still be serviced by the other node connected to it. If two nodes fail, however, there will be connection loss.
- ✓ **Star-ring combinations, or mesh networks:** Mesh networks help provide fault-tolerant access, maintaining a connection if one link is lost.

Redundant Links

Networks strive for redundancy, meaning the use of a backup circuit of some type. This can be achieved in two ways:

- ✓ **A doubled circuit, in which a second, backup circuit still carries some data to the customer if a primary fails:** This sort of doubled circuit can come in two forms: "software redundancy," in which the network detects link loss, and a new link is accessed via software initiation; and "hardware redundancy," in which the two circuits run side-by-side, each delivering a data payload. If one circuit fails, the other circuit simply carries on.
- ✓ **Star-ring, or meshed, connections:** Using a mesh network, each customer access point has more than one connection coming from someplace in the network. This achieves complete redundancy.

Completely redundant physical links are expensive to deploy, and, except when a disconnect is completely unallowable, you won't see them too often. Even when that level of redundancy is necessary, there are other, more efficient ways to achieve circuit redundancy. In networks where users go online with mobile wireless devices, it is more important that network cells have frequent points where they overlap each other (similar to cellphone systems). This allows access to the network through more than one point.

Fault Tolerance via Ancillary Wireless Link

You could back up a wired circuit link by also deploying wireless access. This provides two benefits:

- ✓ At least some data could continue to the customer via the wireless link.
- ✓ The wireless link could help troubleshoot the link loss. Use the wireless link to test data at either end of the link, which helps determine a problem is at the source, at the remote end, or a case of actual cable breakage.

This side-circuit approach could be a bit costly, but in situations requiring absolutely no possibility of disconnect, a wireless side-link could be a lifesaver.

Spectrum Management and Monitoring via Wireless

Although it's not a phenomenon familiar in North America, policing bandwidth is a big deal worldwide. Bandwidth allotment and enforcement is taken very seriously in many quarters. In some arrangements, users license transmission at certain speeds and pay more for faster access. This can work out nicely, because limiting access to higher bandwidth to premium customers insures their speeds will not be impeded by heavy traffic. Providers can offer users a faster-moving, less crowded segment of the spectrum at a premium price. However, in such arrangements, broadcasting or transmitting out of band can cause data delays for your premium clients, interfere with priority data transmission, lead to equipment failure, and just generally create a rather unruly situation for regulators.

Offering bandwidth-specific services, then, could require bandwidth monitoring. What service providers can do is use an ancillary wireless circuit as a bit of a speed cop, watching data rates and transmission on the neighboring wired link. This monitoring would be done by deploying a wireless circuit that runs alongside a wired fiber or copper circuit. The benefits are twofold:

- ✓ A second circuit that is dedicated to monitoring speeds along the main circuit would not burden the main data pathway being sold to customers.
- ✓ Secondly — and this is a huge plus — the second circuit would not be affected by transmission problems in the first. If the wired circuit ever went down, the provider could allow the most premium customers access to the wireless circuit for high-priority transmissions, until the problem is solved.

Summary

You've learned that there's much more to wireless technology than pocket-sized devices. Fixed wireless plays a major role in determining accessibility to voice and data transmission. In the next chapter, we start with your requirements, as an individual, IT technologist, or corporate information officer, and discover what solutions are available for you.

Part III

Sundry Technologies, Products, and Developments

Chapter 12

Big and Little Wireless Solutions

In This Chapter

This chapter lays out common wireless device questions and issues, and offers some solutions. Detailed explanations can be found elsewhere in the book. Specifically, questions regarding individual device purchases, the wireless Web, WLAN access, wireless devices in the enterprise, and application development are addressed.

- ✓ Wireless Web issues
- ✓ Cellphone choices and considerations
- ✓ Pocket PC/PDA issues
- ✓ Wireless device application development considerations
- ✓ Security issues
- ✓ Wireless LAN issues

Wireless Web Issues

This section presents a few questions about connecting to the Web with a wireless device and wireless site development.

- ✓ I'd like to convert my Web site to work with small wireless devices. Any special considerations?

  Because of small screens and slow connections, the wireless Web experience is very different from that on a desktop. Not only are you designing for a fraction of the screen size of a desktop computer, but you also have to keep in mind that the dimensions are different as well. Any benefits of color will be lost on more than half your potential audience, and even the use of simple images will put your site beyond the reach of many devices. Successful small wireless device Web sites impart small amounts of essential information on a topic, and initiate very specific actions, such as ordering a product or requesting information. You may want to consider developing WAP (Web Application Protocol) pages, which utilize a "card and deck" approach (card-sized pages that deliver only essential information or present a single choice), and/or develop Web Clipping applications for the Palm environment.

See Chapter 15 for more information about designing sites specifically for the wireless Web.

- ✓ How can multiple computers in my home share one Internet connection?

  With a wireless router, computers can share an Internet connection through a single computer's Internet access. This computer has an IP address that is exposed to the Internet. The other computers receive internal IP addresses, but can still go online.

- ✓ How do I find out if a particular network service provider covers the area where I live, work, and travel?

  Visit the provider's Web site and look for a coverage map. Service resellers such as OmniSky and GoAmerica offer great deals for PDA and Pocket PC wireless WAN coverage. Check back often, because providers expand coverage areas frequently, as well as the type of services offered.

- ✓ Our company has mobile employees that need always-on access to e-mail and intranet messages, as well as instant notification of events. What's the best solution?

  Probably an advanced paging system such as RIM BlackBerry or the Motorola Adviser Elite. There are many applications for the BlackBerry that provide always-on access to corporate communications and e-mail. See Chapter 1 for more information.

Cellphone Choices

This section discusses points to consider when purchasing a cellphone for individual or corporate use, or as an enterprise solution.

- ✓ I want a cellphone that I can travel abroad with.

  You'll need a Dual Mode or multi-mode phone that will switch between GSM 800 and 1900 MHz frequencies for worldwide access, as well as switch between digital and analog networks. You'll also need a phone capable of switching bands, and network service that facilitates switching. You will be able to make calls out of your immediate network, but it will be expensive. See Chapter 1 for more information.

- ✓ I want a cellphone with a programmable ringer.

  Several brands of phone let you assign ringtones to specific numbers. You'll be able to tell who's calling from the ring alone, without looking at the number. See Chapter 1 for more information.

- ✓ In what situations would a cellphone or hybrid be a good enterprise solution?

  For businesses that require primarily voice-driven applications in the field, where verbal transactions take precedence over complex database entries or remote data access of various types. In such cases, deploying cellphones with PDA capabilities would be a good solution.

- ✓ When picking a cellular network service for enterprise cellphones, what should I keep in mind?

  Make sure the area of service roughly parallels where mobile employees will be spending most of their time, find out if bandwidth offered by the service is sufficient for your needs, and ask about the service provider's plans for 3G service upgrades. Judge whether their upgrade path seems feasible and realistic.

- ✓ Is there a cellular phone operating system best suited for advanced business applications?

  Yes — the Symbian OS, which was developed from the ground up for running communication applications on small wireless devices. See Chapter 4 for more information.

Pocket PC/PDA Issues

This section discusses points to consider when purchasing, accessorizing, or enterprise-equipping a PDA or Pocket PC.

- ✓ What factors should I consider when deciding whether to buy a Pocket PC or a PDA?

  Most PDAs are very light and small, and perform contact management tasks nicely. Furthermore, thousands of applications — for everything from word processing to organizing e-mail — are available for PDAs. For the most part, PDAs have excellent battery life. With the exception of a few devices, such as the Palm VIIx, PDAs require an external modem to access the Internet. PDAs are well suited to the tasks for which they were designed, and can be outfitted with Bar Code scanners and other devices and software for some enterprise applications. See Chapter 2 for more information.

  Pocket PCs run miniaturized versions of Windows applications. They most often have full-color screens and at least 32MB of RAM. Pocket PCs, especially those running the Pocket PC 2002 OS, can handle enterprise applications of many types. Most Pocket PCs require an external modem, LAN PC card, or CF card for online or network access. See Chapter 3 for more information.

- ✓ How do I go wireless with my Pocket PC or PDA?

  That depends on the type of access you want. A WAN card or sled-style modem will provide wireless access from quite a distance. You can access the Internet and a corporate network if the system administrator has installed the protocols and login scripts. WAN cards depend on access through networks such as OmniSky and GoAmerica, or other nationwide service providers. You'll be paying a monthly fee. You can go wireless locally with a LAN PC or CF card, and access a local network. Through the local network, you can access the Internet if configured properly. Also, if you are out on the road and need quick Internet access, you can use a special cable and card and go online via your data-enabled cellphone. You'll have to make sure your cellphone provides data access. If it does, just use a cable/card attachment, such as Ositech's King of Hearts, and dial up your ISP on your cellphone. Using this method, you'll be charged regular cellphone minutes, and the data rate will be on the order of eight or nine Kbps.

- I need a Pocket PC that is fairly rugged and durable — more for work than play.

  Handheld PCs, such as the HP Jornada 760 and NEC MobilePro, have larger screens than Pocket PCs or PDAs. They are "instant on" devices. Just press the On button and you'll be ready to go. Furthermore, they're made to withstand a few bumps. See Chapter 3 for more information.

- I want to print a document while I'm on the road with my Pocket PC or PDA.

  The Pentax PocketJet printer is nearly pocket-sized and prints from most PDAs and Pocket PCs via infrared port. It prints grayscale onto thermal paper. There are no ink cartridges to replace ever. However, printing quality is not impressive. Check out the Canon Bubblejet Portable Color Printer — better quality, but not as portable.

- I'm interested in a device that can audibly navigate directions while I drive.

  With the right software, the newer Global Positioning System units can link with handhelds and laptops to display real-time maps of your current location. Input a destination, and you'll be navigated verbally — told where to turn, and even how to get back on track if wrong turns are made. See Chapter 6 for more information.

- I have an older Pocket PC and would like to install the Pocket PC 2002 OS.

  Some of the Pocket PCs that were released within a year or so prior to the release of Pocket PC 2002 can be upgraded. Some manufacturers offer to do so for a low cost. Pocket PCs must have 16MB Flash ROM to install Pocket PC 2002. At least 32MB of RAM is highly recommended. See Chapter 3 for more information.

- I need to connect my PDA or Pocket PC to a large screen to display presentations and slide shows.

  Hardware products such as Voyager VGA and Presentations To Go will connect your handheld to a big screen via serial port. You can display PowerPoint presentations or similar slide-based shows, as well as mirror any handheld display activity onto the big screen. If your display is large enough, your on-screen actions can be viewable by many users, which is great for teaching groups how to use PDA or Pocket PC applications.

Wireless Device Application Development

Customized applications are being developed for an ever-widening variety of wireless devices. The field is wide open for new programs and ideas. Here are some points to consider:

- I'm wondering if a customized Pocket PC application might move our company forward in a big way. What are some of the specialized applications that can be developed fairly easily?

  Pocket PCs are deployed to perform many tasks — for example, wirelessly extending a network, replacing 2-way radio dispatch, research data collection, point-of-sales applications, and mapping. Microsoft actively supports a developer's kit for the Pocket PC 2002 OS, based on familiar Visual Basic 6 tools. Creating customized applications for the Pocket PC is not too much of a stretch for those familiar with VB.

- I want a phone supported by a large community of developers that can provide application development and many types of software upgrades.

  Java-enabled phones attract the large base of developers familiar with Java. Sun Microsystems has released and actively supports the J2ME (Java 2 Micro Edition), a Java version specifically designed for small devices with limited resources. Owners of Java-enabled phones will be able to enjoy a large variety of powerful applications from many sources.

Wireless Security Issues

Security issues are covered in more detail in Chapter 14. However, the following inquiries may steer you towards the right concerns for your situation.

- I'm concerned about eavesdropping, my phone being cloned, and unauthorized calls being charged to my cellphone.

  Cellular phone users with heightened security concerns should not use analog (AMPS) cellphones and phone services. By nature, digital voice transmission is more secure, as it is easy to encrypt. GSM technology provides very secure voice transmission, and phones based on the Symbian OS (see Chapter 4) are even more secure. Also, discuss encryption and security issues with your service provider.

- What can I do about protecting my wireless device from viruses?

  Major anti-virus software companies such as McAfee and Symantec offer anti-virus software for small wireless devices.

- How can I choose the most secure network for transmitting my sensitive data?

  Pick a CMDA network, which employs spread spectrum technology. Also, if tampering with data on the wireless Web is a concern, consider developing a WAP page, which has a dedicated security layer. If these measures are not sufficient, consider using third-party data encryption services such as those provided by Certicom.

- What about making my WLAN data as secure as possible?

  Utilize Wired Equivalent Privacy (WEP), an optional security feature for the 802.11b spec that provides 40-bit encryption. WEP security is enhanced by thoughtfully distributing WEP keys, (which unlock the encryption) not allowing them to fall into the wrong hands.

- What else can be done to secure wireless data?

  Strict use of passwords, as well as physical tokens such as key cards, will enhance security. Making sure all wireless devices are accounted for will certainly help (not the easiest task when many devices are deployed around the company for mobile use). Also, make security be somebody's job. Designate a person whose task it is to stay abreast of security issues, change everyone's password occasionally, keep anti-virus software current, and watch for security holes in general.

Wireless LAN Issues

This section presents some questions people often have about wireless LAN devices. Chapter 5 discusses WLANs in detail.

- ✓ I'd like to set up a small wireless LAN with just one other computer, without an access point. Can that be done?

  Yes, you can set up a peer-to-peer connection with two wireless LAN cards. They can share file and printer access. With the right software, an Internet connection can be shared if one computer is online.

- ✓ I need to provide network access to wireless mobile devices in a large, multi-structure campus.

  Well-designed wireless LANs can provide network access in multi-floored and multi-building campuses. Multiple access points can be strategically positioned to provide access throughout. Employees or students equipped with wireless devices can enjoy roaming access to network services. Many proprietary WLAN solutions for large campuses are available. The best will allow those who are clustered in areas of high access to use bandwidth from other access points in the system. See Chapter 5 for more information.

- ✓ My company needs portable computers with extra-large displays and LAN access. We need to be able to see large files without scrolling.

  SonicBlue's ProGear is a Web tablet with a 10.4-inch vibrant color screen. Data can be input via touchscreen. It is used for medical and industrial applications, and has a built-in 802.11b connection. See Chapter 5 for more information.

Summary

You've now had a broad look at a number of wireless technologies, applications, and concerns relating to what a user may need. In the next chapter, you get acquainted with a number of diverse wireless technologies not covered earlier.

Chapter 13

Diverse Device Technologies

In This Chapter

As you've learned by now, the world of wireless devices has spawned all sorts of interesting and unusual technologies. This chapter continues to explore such new technologies, looking at hardware, software, and protocols that are on the verge of coming to market (all are well along in their development cycles).

The main groups of diverse device technologies addressed in this chapter are the following:

- ✓ Display technologies
- ✓ Mobile location technologies
- ✓ Mobile power sources
- ✓ Bluetooth
- ✓ Anoto digital paper
- ✓ Synchronization
- ✓ Markup languages
- ✓ Java
- ✓ OBEX

Display Technologies

While traditional display technology for wireless devices, particularly handheld PCs, continues to improve, some potential breakthrough technologies are out there, including the following:

- ✓ Electronic ink
- ✓ OLEDs

Electronic Ink

Imagine a wireless device with a display that looks as crisp and sharp as the piece of paper these words are printed on. Or, imagine a wireless device with a large display that rolls in and out like a window blind, or even folds up for storage. That is the eventual promise that electronic ink holds for wireless devices.

Electronic ink is an ink-like material with three separate components: microcapsules (particle carriers that are about the diameter of a human hair), an inky or oily substance that fills and surrounds the microcapsules, and electrically charged particles that float within the microcapsules. The electronic ink can be printed on a surface, just like normal ink. However, unlike normal ink, the white particles (think of a small ball) within the electronic ink remain free to move within the ink. When an electric charge is applied to a point on a surface covered with electronic ink, the color of that point changes.

This process occurs in two slightly different ways. In the white particle approach, developed by E Ink Corporation, the microcapsules within the electronic ink contain tiny white particles, each of which has a negative charge. The microcapsules are filled with a blue dye. Imagine that the microcapsule is a transparent beach ball, the white particles are Ping-Pong balls, and the beach ball is filled with water dyed blue. Now imagine that you are looking at the clear beach ball from above. Since the Ping-Pong balls float, you will see mostly Ping-Pong balls, which means the ball will appear to be mostly white. If you look at the beach ball from below, you will see the blue-dyed water instead, making the beach ball appear dark. If there were a way to make the Ping-Pong balls float or sink on command, you could change the appearance of the beach ball from light to dark.

Furthermore, if you took thousands of beach balls containing these controllable Ping-Pong balls and spread them across a parking lot, you could write messages that passing airplanes could read simply by commanding the Ping-Pong balls to float or sink. You can do the same thing with electronic ink, except that the "beach balls" are microscopic (100 microns or so in width) and the "parking lot" is a piece of paper or plastic a few inches square.

As mentioned earlier, the white particles in this version of electronic ink carry a negative charge, and therefore, can be moved with the application of an electrical charge. In a wireless device, the electronic ink would be sandwiched between two surfaces in which electrodes are embedded. With these electrodes arranged in an intersecting pattern, it is possible to apply an electric charge to individual points in the electronic ink. The negatively charged white particles move towards a positive charge and away from a negative charge. If a positive charge is applied to a point on the side of the display the user sees, the white particles move in that direction, turning the display white at that point. Applying a negative charge would turn the spot dark. The system driving the display is thus able to draw characters and images. Figure 13-1 depicts a side view of two microcapsules — one where the white particles are drawn to the front of the display by a positive charge, the other where the particles are drawn to the back.

Xerox is also developing electronic ink. Their design involves tiny spheres — rather than white particles and microcapsules. These spheres, which are light on one side and dark on the other, rotate in response to an applied electric field. However, the end result is the same as that of the white particle approach: an ability to change the color of a point in the electronic ink.

The potential benefit of using electronic ink in wireless device displays is enormous. Because the white particles or rotating spheres in the electronic ink only move when an electric charge is applied to them, you don't need to continuously apply power to an electronic ink display. If the image on the display doesn't change, the display draws no power. E Ink, which has been selling an early version of its technology in large display signs since 1999, says that an electronic ink display would consume 50 to 100 times less power than the *liquid crystal displays* (*LCDs*) currently used in most wireless devices. That translates into longer battery life for the device.

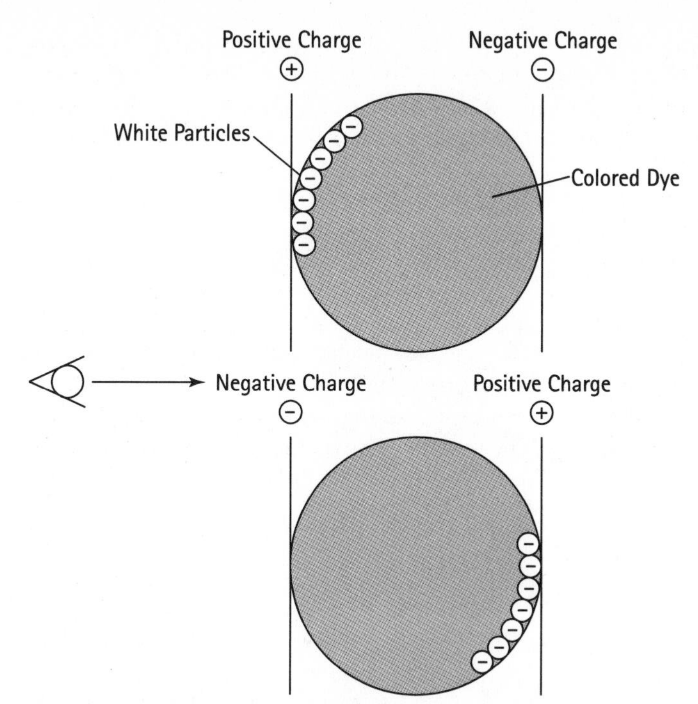

Figure 13-1: Applying the appropriate charge to a point in an electronic ink display changes the color of that point.

Electronic ink has additional advantages. To a high degree, it resembles ink on paper, making displays that use electronic ink easier to read than traditional LCD displays. Furthermore, the reflectivity and contrast are high; the brighter the ambient light, the easier it is to read the display. And with the appropriate electronics to drive them (including perhaps the printable plastic transistors E Ink has licensed from Lucent Technologies), the displays could achieve higher resolutions than current displays at a lower cost. Finally, electronic ink can be printed on flexible surfaces, perhaps leading to roll-up or foldable displays.

E Ink and Philips Components have already announced a black-on-white electronic ink display for handheld devices that has higher contrast than a newspaper and higher reflectivity than an LCD display. Furthermore, it uses ten times less power than an LCD display and will be 30 to 50 percent thinner and lighter. The display has a resolution of 80 pixels per inch — comparable to that available on current handheld devices.

In another interesting development, E Ink has started working with TOPPAN Printing Company Ltd. to develop color electronic ink displays. Although these color displays probably won't appear on the market until at least 2004, wireless devices with monochrome electronic ink displays could be out by 2003.

Currently, the two big players in electronic ink are E Ink Corporation (`www.eink.com`) and Xerox Parc (`www.parc.xerox.com/dhl/projects/gyricon`).

OLEDs

OLEDs, or *Organic Light Emitting Diodes,* are a relatively new display technology that is likely to become popular in mobile devices within the next few years. OLEDs, as their name implies, are made of organic molecules that emit light when a voltage is applied to them. One example of an emissive display technology is a television. Current mobile devices, which usually have liquid crystal displays (LCDs), use reflective display technology. Paper is also a type of reflective display technology.

An LCD, or liquid crystal display, relies on an external light source or a backlight, which reflects off the display.

OLEDs offer significant advantages as a display technology for mobile wireless devices. Because the display emits light, it can be read in bright light, and can also be read without a backlight in low-light conditions. Thus, the display is easier to read. Eliminating the backlight saves power and makes the device lighter. If you've ever seen the brilliant display on the Motorola Timeport phone, you've seen one of the first commercial uses of OLEDs.

OLED MATERIALS AND COMPONENTS

OLEDs typically consist of thin layers of organic material sandwiched between two electrodes. One electrode (the *cathode*) is usually made of metal, and forms the back of the display. Next to the cathode is a layer of material known as the *Electron Transparent Layer (ETL)*. On top of the ETL is the Emitter Layer (the organic material that emits light), then the Hole Transparent Layer (HTL), the Hole Injection Layer (HIL), and the other electrode, or *anode*. The anode is usually a film of Indium Tin Oxide (ITO), a transparent, conductive material deposited on glass or some other transparent substrate. Figure 13-2 illustrates the structure of an OLED.

When a voltage is applied to the OLED, electrons (negative charges) and holes (positive charges) travel through the layers and recombine in the Emitter Layer of the OLED, causing the material to emit light. It typically takes only a few volts (2 to 10) to cause an OLED to emit light.

Color OLEDs can be created by introducing various fluorescent molecules into the Emitter Layer. The Eastman Kodak Company holds the key patents on this technique and licenses them to other companies working on OLEDs.

Glass Substrate	
Indium Tin Oxide (ITO) Layer	ANODE
Hole Injection Layer (HIL)	
Hole Transparent Layer (HTL)	
Emitter Layer	
Electron Transparent Layer	
Metallic layer	CATHODE

Figure 13-2: The structure of a basic Organic Light Emitting Diode (OLED)

Two types of OLED material exist: small-molecule and large-molecule (polymer). Small-molecule OLEDs are created by depositing the layers using vapor sublimation in a vacuum chamber. Polymer OLEDs can be created using solvent coating techniques. However, a more interesting approach to creating polymer OLEDS was developed in 2001 by Philips Research, a unit of Royal Philips Electronics. Philips Research developed a high-resolution inkjet printing technology suitable for industrial-scale production of color passive-matrix OLEDs. In other words, they came up with a way to *print* color OLED displays.

OLED DISPLAY TYPES

There are two display designs for OLEDs: the passive-matrix design mentioned in the previous paragraph, and an active-matrix design. The passive-matrix design consists simply of an array of OLED pixels, connected by intersecting anode and cathode conductors. External circuits drive each row and column (anode and cathode, or vice versa). A current is applied to each row (or column) sequentially. When the current is applied to a row, a data signal is applied to each column (or row) that contains a pixel to be turned on in that row. The entire array of OLED pixels gets scanned perhaps 60 times a second to create the image on the display. In general, passive-matrix OLED displays are well suited for low-cost alphanumeric displays and other low-resolution uses.

Active-matrix OLED displays integrate driver electronics onto the substrate for each OLED pixel. In an active-matrix OLED display, the display system can address each pixel independently. There is no theoretical limit to the number of pixels or the resolution of an active-matrix OLED, which means that an active-matrix OLED could potentially be made any size, with any arbitrary resolution. In general, active-matrix OLED displays are most beneficial when high-resolution and using information are important considerations — in video and graphic displays, for example.

OLED APPLICATION

Thanks to the materials that comprise OLEDs, and the way they are made, OLEDs have some interesting potential uses. A company called Universal Display Corporation recently signed a contract with the Defense Advanced Research Projects Agency (DARPA) to develop a type of OLED display known as a *Flexible Organic Light Emitting Diode*, or *FOLED*. FOLEDs would be bendable, perhaps even to the point where they could be folded up and stored in a pocket when not in use.

Universal Display Corporation is working on *TOLEDs*, or *Transparent Organic Light Emitting Diodes*. TOLEDs would be extremely useful for applications that need to deliver visual data to the user without completely obstructing the user's view. Examples include displays that mount on the lenses of a pair of glasses, or are applied to the windscreen in the cockpit of an aircraft.

For more information on OLEDs, visit the following Web sites: www.kodak.com/global/en/professional/products/specialProducts/OEL/oelIndex.jhtml (the Eastman Kodak Corp. OLED site), and www.universaldisplay.com/tech.php (the Universal Display Corporation site).

Mobile Location Technologies

There is great value in mobile location technology, which enables you to determine exactly where a mobile device (and the person using it) is located. And this type of technology doesn't just benefit marketers looking for ways to sell a person something at the nearest store. The U.S. Federal Communications Commission (FCC) also recognizes the advantage of mobile location technology in crisis situations — if mobile phone companies could quickly pinpoint the location of a phone, the response time of emergency services would increase.

The FCC maintains a Web site on Enhanced 911, which provides the rules being imposed (these rules govern the reliability and availability of E-911 service), relevant press releases, links to related sites, and a wealth of other information The site is found at www.fcc.gov/e911/. Note that as of this writing, wireless phone companies are having difficulty complying with the rules proposed by the FCC, and have petitioned the government for more time to come into compliance with those rules.

You can divide mobile location technology into two types of systems: terminal-based and network-based. Terminal-based systems require some sort of position-processing power in the wireless device, or a SIM card. GPS is an example of a terminal-based system. Network-based systems put the position-processing power in the wireless network itself, and do not require any special processing power in the device or SIM. The Enhanced 911 system that would be used for pinpointing emergency calls is an example of a network-based system. Hybrid systems that combine network-based and terminal-based processing are also possibilities. A-GPS is an example of a hybrid system.

Each approach to determining location has benefits and drawbacks. Terminal-based systems require new mobile terminals, or Subscriber Identity Modules (SIMs), that are capable of performing the location computations themselves. Since new terminals are needed in terminal-based systems, full compliance with the system doesn't occur until all users have replaced their terminals (or SIMs). The primary advantage of terminal-based systems is that upgrading to the new terminals is the responsibility of the user. The network operator doesn't need to spend vast sums of money to upgrade the network; users pay for the capability if they want or need it.

The ubiquity of network-based systems, on the other hand, is in the control of the network operator, since such systems require no assistance from the terminals or SIMs. Thus, operators can quickly achieve 100 percent coverage within their networks by deploying the needed equipment whenever they are ready.

Expert Advice: Protecting Your Location Information

In the United States, the idea that big corporations or government could determine an individual's whereabouts by tracking their mobile device scares many people. For one, many Americans have a strong concern for privacy. They simply don't want their location known and monitored. Also, they question the appropriate use of such information. Most people like the idea that emergency services can find them when they need help, but they object to the fact that marketing organizations could send ads for the nearest lingerie store to their mobile phone. This situation has prompted political action.

In July 2001, Senator John Edwards, a Democrat from North Carolina, introduced a bill to the United States Senate that would allow users of wireless devices to elect to prevent their position from being monitored through their wireless device. If passed, this legislation would regulate when people could be tracked — and also how the information gained by tracking them could be used.

Although the use of wireless devices is just becoming widespread, it appears that a consensus is forming on the appropriate way to control location information. This approach, called *opt-in*, is similar to the approach being advocated for many other privacy-related issues, such as the sale and use of personal credit records. With an opt-in approach, the user must actively choose to make location information available at any given time.

An opt-in approach to mobile location information would likely entail adding a control to wireless devices that, when activated, allows the device to transmit position information, or signals the wireless network that it is now "okay" to calculate the mobile terminal's position.

While opt-in location systems have some drawbacks (the cost of adding the capability and the possibility that the user will not be able to activate the system during emergency situations), they seem to be a decent compromise between privacy issues and the benefits of mobile location.

Wireless privacy is a hot issue in Washington, D.C., right now. Since the number of wireless users is still relatively small (compared to what it may be in 5 to 10 years), that means that anyone reading this book has a real opportunity to influence the debate and the final solution. To make your opinion heard, start by contacting your political representatives. Then contact your wireless carrier, the Center for Democracy and Technology (CDT) (`www.cdt.org`), and the Electronic Privacy Information Center (EPIC) (`www.epic.org/`). Finally, contact the Chairman of the Federal Communications Commission (FCC). See the FCC Web site contact page (`www.fcc.gov/contacts.html`) for the appropriate e-mail address.

Hybrid systems have the slow deployment characteristics of terminal-based systems, along with the costs of network-based systems, but they do offer the potential for the most accurate position calculations of any approach.

Terminal-Based Position Systems

In terminal-based position systems, the terminal calculates its own location. That doesn't mean the terminal makes the calculation without outside assistance; however, the terminal does use whatever data it can to perform the calculations within the device.

The prototypical terminal-based system is GPS. The *Global Positioning System* (*GPS*) is a system that allows devices equipped with the appropriate receivers to determine their position (latitude and longitude) and altitude, as well as the exact time from virtually anywhere on the earth, or close to its surface.

The best source of GPS information online is the GPS site maintained by the United States Federal Aviation Administration (FAA). It is located at `http://gps.faa.gov/`.

GPS is built around a constellation of 24 satellites in orbit around the earth at an altitude of 11,000 miles. These satellites continuously transmit their position, along with an extremely accurate time signal. By measuring how long it takes for this signal to travel from the satellite to the receiver, the data terminal can calculate its distance from the satellite.

If a receiver can receive signals from four satellites simultaneously, it is possible to compute the latitude, longitude, and altitude of the receiver, along with the exact time. The data terminal or other device containing the receiver accomplishes this feat by triangulating its position using the information obtained from three of the four satellites. The information from the fourth satellite provides time correction information. The resulting calculations can determine the location of the receiver to within 100 meters, 95 percent of the time, anywhere on earth.

One hundred meters is 300 feet, so in this basic mode of operation, the receiver can be anywhere within 300 feet of the location shown by the receiver. Some of this potential error is due to the physics of the situation, with particles in the atmosphere and reflections off macroscopic objects (buildings, hills, and so on) causing small variations in the amount of time a signal takes to travel from the satellite to the receiver.

A larger error is caused by something called *Selective Availability* (*SA*). The United States military can turn on SA to induce slight errors in the data from satellites, thereby increasing the size of the errors in the GPS calculations. Fortunately, SA has been disabled since May of 2000, allowing unaided GPS units to achieve significantly greater accuracies. With SA disabled, the typical error is around 30 meters (100 feet).

Thirty meters is an accurate location, but narrowing it down still further would be helpful. Fortunately, there is a way to do this. It involves positioning GPS transmitters on the ground at precisely known locations. These stations can effectively remove most of the error from the GPS position calculations in their area. If these error corrections are made available to mobile GPS receivers, they can achieve accuracies of less than 10 meters in the vicinity of the ground stations. See the sidebar "Augmenting GPS with WAAS" for one method of getting the error corrections to mobile GPS receivers.

Augmenting GPS with WAAS

WAAS, or *Wide Area Augmentation System*, is a technology that can greatly improve the accuracy of civilian GPS location calculations. WAAS was developed by the Federal Aviation Administration (FAA) to augment, or improve, the accuracy of GPS signals by calculating the errors in GPS signals and transmitting corrections to those errors.

The WAAS system uses several ground stations that measure the signals from GPS satellites. The stations then determine the errors in those signals and compute corrections for them. WAAS transmits the corrections to several satellites in geosynchronous orbit. Those satellites, in turn, broadcast the corrections to GPS receivers. The corrections are broadcast using the same frequency as standard GPS signals, so any GPS unit can receive them. Units that have the proper software can use the error corrections to improve the accuracy of the signals from normal GPS satellites, thus improving the accuracy of the final position calculation.

The FAA reports that GPS receivers should be able to achieve six- to seven-meter accuracy when using WAAS. However, Magellan, a maker of handheld GPS units, reports that they are typically seeing accuracies of around three meters during testing of their own equipment.

WAAS has some drawbacks. Compared to unaugmented GPS, WAAS satellites are geosynchronous, which requires them to orbit above the equator. The satellites may not be visible in certain locations, depending on the local geography. The problem is worse, of course, the further north or south you go. Also, as long as the system is under development (and as long as it is limited to only a few transmitting satellites), the system could go offline for a period of time.

Despite its imperfections, WAAS looks like a real advance for GPS users. When buying GPS, check for WAAS compatibility, which entails enough memory and the right software to use the correction signals.

The FAA maintains a public Web site about the WAAS program. It is located at `http://gps.faa.gov/Programs/WAAS/waas.htm`.

Network-Based Position Systems

As of this writing, a variety of network-based position systems are in development or in use. The following sections cover two of these systems: CGI+TA and UL-TOA.

CGI+TA

CGI+TA stands for *Cell Global Identity and Timing Advance*. This GSM system uses the knowledge of which cell a terminal is in, combined with the timing advance, to determine the approximate location of the terminal within the cell. The cell the terminal is located in is determined by the cell global identity (CGI) of the base transceiver station (BTS) communicating with the mobile terminal. The CGI is unique within the GSM system, and can be used to determine the exact geographic location of the BTS. The timing advance is the delay between the beginning of a time slot and the arrival of data from the mobile terminal at the BTS. The timing advance allows the network to estimate the distance of the mobile terminal from the BTS with an accuracy of approximately 550 meters.

Knowing which cell a terminal is in, and measuring the distance of the mobile terminal from a single BTS serving that cell, only makes it possible to determine the position of the mobile terminal along an arc, as shown in Figure 13-3.

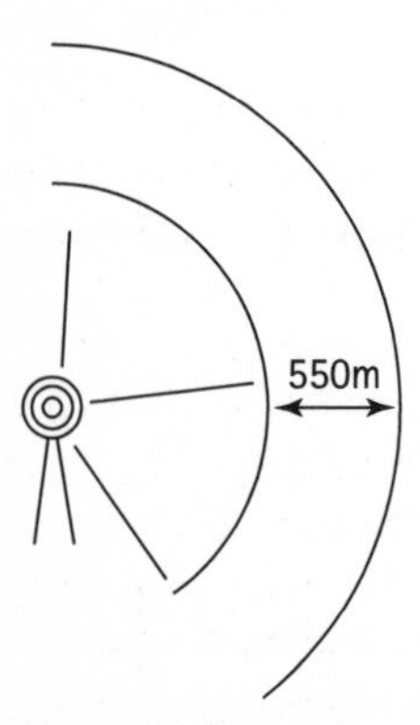

Figure 13-3: Knowing cell terminal and base transceiver identity as well as station–terminal distance makes position calculation possible.

Given that the terminal is located somewhere within the arc determined by the cell it is in and the distance to the BTS, the overall accuracy of the calculated position depends on the size and shape of the cell. If the cells of the network are laid out in a hexagonal grid, the mobile terminal can lie anywhere along a 120-degree wide arc. This means that the further the terminal is from the BTS, the longer the arc is, and the less accurate the position calculation is.

UL-TOA

In the *Uplink Time of Arrival (UL-TOA)* system, the position of the mobile terminal is computed by measuring the time of arrival of a signal from the mobile terminal to four or more Location Measurement Units (LMUs). Each LMU measures the time of arrival of bursts of data from the mobile terminal, then passes this information to the *Mobile Position Center* (*MPC*).

The MPC computes the *Time Difference of Arrival* (*TDOA*) for pairs of LMUs. The MPC uses this information, along with the known geographic location of each LMU and the timing offset between the LMUs, to compute the position of the mobile terminal.

The result of the MPC's calculations is a position estimate, along with an uncertainty estimate for each mobile terminal. The accuracy of the calculations is affected by the environment in which the mobile terminal is operating, along with the number of LMUs involved. Position estimates using UL-TOA can be accurate to within 50 meters in rural environments and to within 150 meters in urban environments.

Hybrid Position Systems

A-GPS and E-OTD are two examples of hybrid position systems.

A-GPS

Assisted GPS (A-GPS) is a hybrid mobile positioning system that supplements standard GPS signals with signals from Location Measuring Units (LMUs) that are integrated into the wireless network.

E-OTD

E-OTD stands for *Enhanced Observed Time Difference*. In this hybrid system, the mobile terminal measures the time difference of the arrival of signals from nearby base transceiver stations. These signals can be synchronization signals, or any other easily identifiable elements of a normal transmission between the mobile terminal and a BTS.

Whatever element of the BTS signal is used, the terminal must measure the time difference in the arrival of that signal from nearby pairs of BTS — known as the *observed time difference (OTD)*. However, since BTS signals are not normally synchronized, this information by itself is of little use and must be enhanced with the *relative time difference (RTD)* for the same BTS pair.

With the OTD and RTD values for at least three pairs on BTS, as well as the exact geographic coordinates for each BTS, it is possible to accurately triangulate the position of the mobile terminal. The calculations necessary to do this can be performed by the mobile terminal or by the wireless network.

No matter which part of the system does the calculations, the result is a calculated position for the mobile terminal. E-OTD systems can reportedly deliver positions that are accurate with 60 to 200 meters, depending on local conditions.

Mobile Power Sources

Mobile wireless devices need mobile power sources. And given our desire to make our mobile wireless devices ever smaller, ever lighter, ever faster, and ever more capable, we need mobile power sources that provide more and more power for longer and longer periods of time. Lithium ion batteries, which have been the power source of choice for mobile devices of all sorts, are not keeping up with the demands we are placing on them. Several companies are trying to resolve this problem by developing new kinds of batteries or by replacing traditional batteries with new technologies, such as:

- ✓ Zinc-air batteries
- ✓ Fuel cells

Zinc-Air Batteries

Zinc-air batteries, which have been around since the 1800s, are the power source of choice for hearing aids. However, they are now gaining attention as a mobile power source for wireless devices, despite some quirks that would at first seem to make them impractical for this type of use. For example, zinc-air batteries are *primary cells*, which means that they are not rechargeable. You use them once, then dispose of them. This goes counter to the direction of most mobile device batteries, which are *secondary cells* (they are rechargeable). In addition, zinc-air batteries require oxygen from the atmosphere to function.

In spite of these quirks, zinc-air batteries offer a major benefit for use in mobile wireless devices. Every type of battery technology can store a certain amount of energy in a given volume, which is known as its *energy density*. Zinc-air batteries use oxygen from the atmosphere to achieve an energy density that is several times the energy density of the lithium ion batteries currently used in most mobile devices. They work by controlled oxidation of the metal zinc, in a manner similar to standard alkaline batteries. The difference is that alkaline batteries store the oxygen they need within the cell, in the form of manganese dioxide. Since zinc-air batteries get their oxygen from the atmosphere, they can include more zinc and deliver more power.

Internally, a standard zinc-air battery consists of a zinc anode, a conductive potassium hydroxide electrolyte, and oxygen from the atmosphere, which acts as the cathode of the battery, and passes into the cell through holes in the casing. Figure 13-4 depicts a schematic diagram of such a cell.

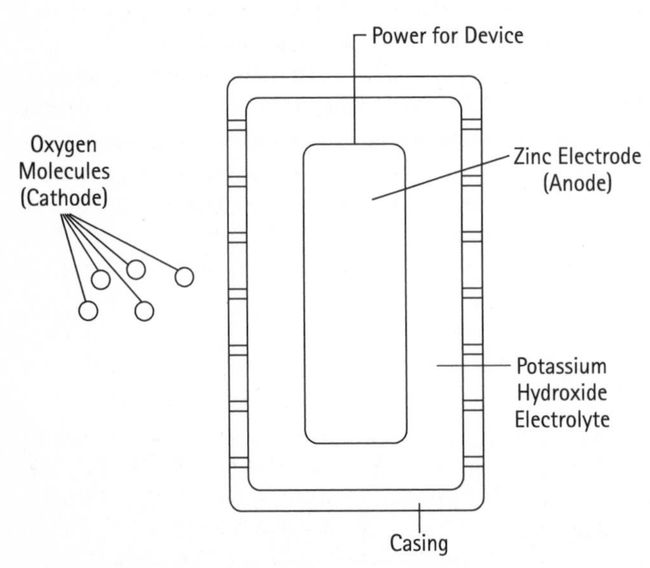

Figure 13-4: The structure of a basic zinc-air battery

One leading supplier of zinc-air batteries for wireless devices is Electric Fuel Corporation. They produce an Instant Power Battery series, which is sold as an alternative to conventional cellphone batteries. Their batteries, for example, produce three to five times as much power as the rechargeable batteries in a typical cellphone. Since zinc-air batteries typically lose only about 2 percent of their power when stored for a year in their original airtight container, they are ideal for use as backup batteries. Once exposed to the atmosphere, the batteries start to discharge and will be almost fully discharged within a few months.

At least one company (AER Energy Resources) has created rechargeable zinc-air batteries that can be recharged about 30 times before failing, greatly increasing the usable life of a cell.

Zinc-air batteries are environment-friendly, as they contain no heavy metals or other hazardous compounds. In addition, they can be recycled easily.

The Electric Fuel Corporation Web site can be found at `www.electric-fuel.com`. The AER Energy Resources Web site can be found at `www.aern.com`.

Fuel Cells

Fuel cells are power sources that run on two elements: hydrogen or a fuel high in hydrogen, and oxygen or a source of oxygen, such as the atmosphere. As long as the hydrogen fuel enters the cell and oxygen is available, the cell generates electricity. Fuel cells offer several advantages over conventional batteries, with each fuel cell design reflecting its own unique advantages.

At their most basic, fuel cells are similar to batteries. That is, they have two electrodes with an electrolyte in the middle. But that's where the similarities end. In a fuel cell, the electrolyte is usually a polymer membrane of some sort, instead of a paste like those found in batteries. When a fuel cell generates electricity, hydrogen or a hydrogen-containing fuel such as methane passes over the anode of the cell. There, the hydrogen atoms give up an electron and quickly pass through the electrolyte (membrane).

Once through the electrolyte, the positively charged hydrogen atoms reach the cathode, where they encounter electrons (the same as those stripped from the hydrogen atoms on the other side of the membrane) and oxygen. They travel to the cathode through an electrical conductor, producing a current that can be used to power a wireless device. The positively charged hydrogen atoms, electrons, and oxygen then combine at the cathode to create water. Figure 13-5 shows this process schematically.

There are several variations on the basic fuel cell design. Most designs use a catalyst (such as platinum) on the anode to strip the electrons from the hydrogen atoms. Platinum is effective, but it is also expensive. Furthermore, it can pose an environmental problem when the fuel cell is discarded. Some fuel cells, such as those created by PowerZyme, work without a metallic catalyst, instead using methanol fuel and a sophisticated membrane. These fuel cells have no moving parts, are environmentally-friendly, and feature a power density that is several times that of zinc-air batteries.

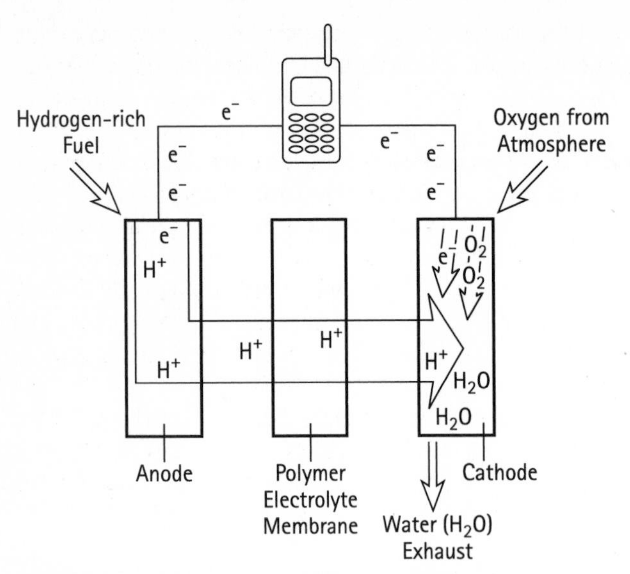

Figure 13-5: A diagram showing the basics of fuel cell operation

Manhattan Scientifics is developing the MicroFuel Cell™, which will hopefully provide ten times the energy density of lithium-ion cells. The theoretical maximum appears to be about 30 times the energy density of lithium-ion cells. PowerZyme expects to have its Gen-One first generation fuel cells on the market in 2002, with full-scale production by 2003. Manhattan Scientifics expects to be beta testing their fuel cells in 2002.

The Manhattan Scientifics Web site can be found at `www.mhtx.com`. PowerZyme doesn't appear to have a publicly accessible Web site at this time.

Bluetooth

Bluetooth is the name of a short-range, low-power radio frequency wireless technology that has recently been gaining widespread attention. (No, we're not going to tell the story of how Viking King Harald Bluetooth united Denmark and Norway. You've probably heard it already, and it isn't that exciting anyway.) Bluetooth is similar to wireless networking technologies such as 802.11, but it focuses on ad hoc connections between nearby PDAs, PCs, phones, printers, keyboards and the like, ideally replacing the meters of cable that snake through most offices.

The vision for Bluetooth-equipped devices includes scenarios like this: You arrive at your office early in the morning, your wireless data terminal in your briefcase. As you enter the building, the data terminal and the wireless systems in the building begin communicating with each other through the low-cost, short-range Bluetooth transceivers embedded in each. The data terminal sends the high-speed color printer on your floor a copy of the proposal you edited at home last night, and tells it to print three copies. Meanwhile, the company network tells your data terminal that your boss has scheduled a meeting in 45 minutes. Synchronizing the copy of your calendar in the data terminal with the one on the main server in the office, the terminal notes a conflict and alerts you. While this is happening, you take the data terminal out of the briefcase and connect it to your monitor (the data rate for Bluetooth is too slow to be practical for driving the display). The keyboard and mouse also contain Bluetooth transceivers, so you don't need to physically connect them to the data terminal. When the conflict notification message appears on your monitor, you realize you'll have to skip your coffee break today, so you update your calendar. The terminal sends the change to the main server (using Bluetooth, of course).

The preceding example illustrates some of the uses envisioned for Bluetooth devices. In fact, the designers of Bluetooth formulated a number of *Bluetooth usage models*, each of which defines a specific way a user can interact with a Bluetooth device. For the initial versions of Bluetooth, the designers specified five usage models:

- ✓ **The Three-in-One Phone:** A phone that uses the best telecommunication technology for any given situation. When it is near another Bluetooth-equipped phone, it functions as a walkie-talkie, thereby avoiding all wireless communications charges. When at home, the phone uses Bluetooth to connect wirelessly to your home phone system, thereby only incurring wired phone charges. When neither of these situations apply, the phone functions as a normal mobile phone.
- ✓ **The Internet Bridge:** The user connects to the Internet using a Bluetooth link to a Bluetooth-equipped mobile phone, or an adapter that is connected directly to the wired phone system.
- ✓ **The Interactive Conference:** During a meeting, participants can use Bluetooth to wirelessly transfer files to selected recipients, as well as electronically exchange business cards.
- ✓ **The Ultimate Headset:** A Bluetooth-equipped headset that automatically makes wireless connections to a mobile phone or an adapter that connects directly to the wired phone system, allowing the user to concentrate on more important tasks.
- ✓ **The Automatic Synchronizer:** Allows any Bluetooth-equipped device that also has *PIM* capabilities to update that data to agree with the same data in other devices that belong to the same user. In the scenario described earlier, the Automatic Synchronizer usage model covers the interaction between the data terminal and the main server with regard to the new meeting request.

Many in the industry focus on Bluetooth's potential to create a wireless personal area network (wireless PAN), a spherical area perhaps ten meters in diameter (called a Personal Operating Space, or POS) within which your personal Bluetooth devices form a wireless network. If such wireless PANs existed, and if workplaces were designed with wireless PANs in mind, you could walk into a workspace and immediately take control of the space as if it were your own. Outside the workspace, your wireless PAN could continuously search for systems and devices to communicate with. For example, you might read your e-mail while sitting in the back of a Bluetooth-equipped taxicab. Or you might receive notice of a flight change while walking past an access point in the airport. You could also walk past a store and receive an instant message, notifying you of their latest sale.

Now that you know what Bluetooth is and how it might be used, it is time to learn a bit about how it works. Bluetooth is a wireless RF technology that operates in the 2.4 GHz (2,400 MHz) *ISM* band of the spectrum — the same part of the spectrum used by 802.11 wireless networks, microwave ovens, some newer cordless telephones, and assorted other devices. The ISM band comprises 79 1-MHz channels (2.402 GHz, 2.403 GHz, up to 2.48 GHz), all of which are available for unlicensed use.

With so many potential users of the same spectrum, devices that operate in the ISM band must be able to handle interference and minimize the interference they cause by employing a spread spectrum air interface. Bluetooth devices, for example, use *Frequency Hopping Spread Spectrum* (*FHSS*) technology for their air interfaces. The Bluetooth implementation of FHSS in this band is designed to maximize usable bandwidth within each channel, minimize interference, and enable low-power operation.

In most parts of the world, all 79 channels in the 2.4 GHz ISM band are available for use in the hopping pattern employed by Bluetooth devices. To accommodate areas where these channels are not, Bluetooth transceivers also support a hopping pattern that uses 23 channels. With the transmitters hopping to new frequencies hundreds or thousands of times a second, it is vital that the transmitter and receiver both use the same hopping pattern. To understand how frequency hopping works specifically in Bluetooth, you need to understand how Bluetooth devices communicate with each other.

Understanding Frequency Hopping Spread Spectrum

Two common forms of spread spectrum technology exist: Frequency Hopping Spread Spectrum (FHSS) and Direct Sequence Spread Spectrum (DSSS). While DSSS uses a spreading signal to literally spread the data signal across a much wider channel than it would otherwise occupy, FHSS broadcasts a single data packet on a channel, then changes frequency (hops) to a new channel to send the next data packet. The rate at which the signal hops to a new channel depends on the system transmitting it. For Bluetooth, the signal hops 1,600 times a second.

FHSS satisfies the requirements of minimal interference with other devices and the ability to handle interference from other devices. Since only a single data packet is transmitted on a given channel at any time, only that packet can interfere with other devices using the same channel at the same time. And since systems transmitting on these channels use FHSS, the chance that the devices will hop to the same channel and interfere with each other on their next data packet is extremely low. Interference between any two devices is intermittent, infrequent, and of short duration.

To handle interference when it does occur, receivers can request that a data packet be retransmitted if it was not received properly. By this time, the transmitter and receiver are on a different channel than the one on which the interference occurred, so odds are good that the packet will get through when retransmitted.

Two to eight Bluetooth devices communicating with each other form a *piconet,* or shared communication channel through which members of the piconet communicate.

The wireless PANs described earlier in this section are a form of piconet.

One device in the piconet assumes the role of master, while all other devices in the piconet are slaves. The master device defines numerous features of the piconet, including:

- ✓ The frequency hopping pattern to be used in the piconet.
- ✓ The timing of the frequency hops.
- ✓ Which frequency the piconet is currently using.
- ✓ Which slave device can transmit or receive next.

Multiple piconets existing in the same area at the same time are called scatternets.

The last item in this list (which slave can transmit or receive next) highlights another important feature of Bluetooth technology. Bluetooth devices are polled by the master — they cannot transmit whenever they wish.

The speed at which any particular Bluetooth device moves data over the air interface is constrained by the piconet it is currently part of. While the raw data rate of a Bluetooth device (which is limited by the channel size and by government regulations) is one Mbps, the actual maximum full-duplex (two-way) data rate of a Bluetooth device is somewhere around 430 Kbps. Since all Bluetooth devices in the piconet use the same hopping pattern and are transmitting on, or listening to, the same channel, they must share that channel. Thus, the larger the piconet, the lower the effective data rate for each device.

Security is an important aspect of a wireless technology that is used to transfer personal information. Because Bluetooth is a spread spectrum technology (which is, by nature, harder to intercept than other wireless technologies), it has inherent security advantages. In addition, Bluetooth devices encrypt the data traveling over the air interface. The encryption scheme, described in detail in the Bluetooth specification, defines encryption keys of up to 128 bits in length, although certain governments limit the length of the encryption key to ensure that their own security and intelligence forces can break the code if necessary.

With the basics of Bluetooth technology now covered, you should have enough information to decide if you want delve further into the subject.

Expert Advice: Waiting for Bluetooth

The primary concern about Bluetooth is whether it will work in the real world. We think it will — eventually. Some kinks still need to be worked out, but Bluetooth has already been demonstrated in the lab, and there are some Bluetooth devices already on the market.

As of mid-2001, numerous vendors have Bluetooth devices that work and communicate with other Bluetooth devices *from the same manufacturer.* Interoperability of devices from different manufacturers has been spotty at best. Attempts to set up ad hoc piconets at Bluetooth UnPlugFests (industry meetings where representatives from manufacturers come together to see if their Bluetooth devices will work together) have been less than unqualified successes.

That said, there are still many good reasons to get Bluetooth working properly, not the least of which is the need for a technology that gives us the connectivity of cables without the expense and inflexibility of a standard wireless LAN. In addition, many people find the concepts of wireless PANs and POSs compelling. Snags are common in technology development efforts, and furthermore, companies invest many person-years of research and development in a new technology. We believe that Bluetooth is coming, and it will be successful. It just may take longer and be less earthshaking than its most ardent proponents would have us believe.

For the most complete information on Bluetooth (albeit information with a pro-Bluetooth slant), visit the Official Bluetooth Web site at `www.bluetooth.com/default.asp`. It provides the Bluetooth specification, white papers, usage models, and the Bluetooth SIG.

Anoto Digital Paper

What if you could send whatever you wrote to any computer in the world? A Swedish company called Anoto AB has developed a new technology that enables individuals to write with Bluetooth-equipped digital pens on communications-friendly digital paper. With Anoto functionality (defined in the *Anoto standard*), the combination of digital pen and paper can produce situations like these:

- ✓ In the middle of a meeting, you are struck by a brilliant idea. Using a digital pen, you scribble it down in your notebook, which (of course) uses digital paper. When you are finished writing, you check a small Send box at the bottom of the page. Instantly, the pen transmits your notes to your Bluetooth-equipped wireless data terminal, which then forwards the information to your desktop PC in the next building over.

- ✓ Driving to work, you get a flat tire. After consulting your digital paper–based personal organizer, you decide to block off a couple of hours in your morning schedule for dealing with the car. You mark the time as busy, then check the Update box. Your digital pen uses Bluetooth to contact your phone, which connects to the Internet, which contacts the servers at the office and tells them to update your online calendar.
- ✓ As you travel through the airport on the way home from your dream vacation, you realize that you forgot to send your parents a postcard. You duck into the nearest gift shop and buy a postcard printed on digital paper. You scribble a note to your folks, enter their e-mail address in the space provided, and check Send. Your digital pen contacts the nearest airport Bluetooth wireless access point, and your postcard is sent across the Net to your parents' house.

Anoto AB has licensed its technology to major technology and paper companies around the world. The special Anoto paper and pens should be on the market before the end of 2001.

How exactly do you turn an ordinary pen and paper into a Bluetooth-equipped digital pen and communications-friendly digital paper? Imagine, if you will, a huge piece of paper with a surface area of millions of square kilometers. Suppose someone prints a very fine pattern of dots on this paper. Some of the dots are slightly displaced from a perfect grid pattern, with the displacements such that they do not repeat on the entire, multicontinent piece of paper. This pattern is known as the *Anoto pattern*. You only have to look at a tiny portion of the Anoto pattern to identify your exact location on the "megasheet." Now imagine chopping the megasheet into billions of normal-sized sheets. If you look at the Anoto pattern on any one of these smaller sheets, and know how to interpret the pattern, you can uniquely identify every point on every sheet by its location in the original megasheet.

Finally, imagine that your Anoto digital pen can not only write on paper using ink, it can read the Anoto pattern on a piece of paper and record every point the pen writes on. When you check the appropriate box on the piece of paper, the pen transmits a record of what you wrote to a nearby Bluetooth device. From there, the information travels through the Internet to any destination computer or wireless device in the world.

Fortunately, Anoto digital pens and digital paper are not just imaginary. By the time you read this chapter, they should be available in stores. Anoto AB already has partnerships with heavyweight companies such as 3M, A.T. Cross Company (the maker of Cross pens), Ericsson Mobile Communications, Mitsubishi, Montblanc, and Pilot (the pen company). In conclusion, we think this is one technology that is going to make it off the drawing board and into your office.

For more information on Anoto AB and the Anoto standard, visit the Anoto Web site at `www.anoto.com`.

Synchronization

Updating your personal data — contacts, calendar, and the like — is relatively easy if you keep this data in only one location, say, your office computer system. You need only enter the data into your *Personal Information Manager* (*PIM*) software and keep it up-to-date. However, once you add a mobile device (with its own PIM) to the picture, you run into problems. For example, how do you ensure that the data in the mobile device matches the data in your office computer system? Imagine this situation: You have a desktop PC in the office (connected to the corporate groupware applications) and another at home, a laptop that travels with you, a RIM BlackBerry pager, a Palm V, and a Psion 5mx. Granted, this may be an extreme case, but think how many places you could keep copies of your personal data if only you knew how to keep them all up-to-date.

Synchronization and copying data between devices can be a problem because of compatibility and interoperability issues. It can be hard to plan or predict when conflicts will occur, however. For example, consider this situation: You are at a customer's office and you need to make an appointment for a follow-up meeting. You pull out your trusty Palm OS device and add the new appointment to your calendar before heading back to your office. However, while you are out your boss decides it is time to give you that big bonus you deserve. He sets up a meeting that overlaps with the follow-up meeting you scheduled. Your boss uses the corporate groupware application to enter the meeting into the copy of your calendar that is on the corporate network. What happens when you return to the office? If you copy your calendar from the corporate network onto your Palm OS device, you'll make it to the meeting with your boss, but you'll miss your follow-up meeting with the client. If, on the other hand, you copy your calendar from your Palm OS device to the corporate network, you'll make it to that follow-up meeting, but you'll miss the one with your boss.

The answer to this problem is, of course, to use synchronization software. Synchronization software doesn't simply overwrite the personal data in one place with the data from another place. Instead, it analyzes the data at various levels of detail to determine what it should keep, and then puts that data on both systems.

A company called Pumatech, Inc. is the leader in this arena. Pumatech produces the Intellisync line of synchronization software, which supports more kinds of devices (Palm OS, Pocket PC, Windows CE, and Symbian) and more PIM software than any other product on the market. Pumatech's desktop products synchronize between mobile devices and desktop PIMs such as Lotus Notes, and their server-based products synchronize between mobile devices and groupware applications such as Microsoft Exchange.

Not only does Pumatech produce the widest array of synchronization products, and synchronize the most devices and applications, the company is actively setting the standards for future wireless synchronization applications. Pumatech worked with the Bluetooth people to help craft the synchronization features of Bluetooth. It is also a member of the group crafting SyncML, a synchronization-specific language that could soon be supported by many wireless devices.

For more information on SyncML, see the "Markup Languages" section of this chapter. Bluetooth technology is covered in the "Bluetooth" section of this chapter.

To provide the most effective synchronization, Intellisync examines the individual fields within the personal data records that are being synchronized. This gives the program the best odds of determining which data to keep. Still, the software may not be able to resolve certain conflicts (two meetings scheduled at the same time, for example). In such cases, users can specify which system will override the other, or can have the software notify them of the conflict and let them apply human reasoning and judgment — the bonus or the client?

Anyone who uses more than one system that contains personal information can benefit from synchronization software, and would be well advised to investigate the possibilities.

The Pumatech, Inc. Web site can be found at `www.pumatech.com`.

Markup Languages

The American Heritage Dictionary of the English Language defines a *markup language* as "a coding system, such as HTML and SGML, used to structure, index, and link text files." Markup languages don't sound like they would be of high importance to wireless devices, but they are. To be convinced of this, you need only realize that *HTML*, the *HyperText Markup Language*, is the language in which standard Web pages are written. What follows is some general information on markup languages, followed by a short description of five markup languages that are of particular interest to users of wireless devices:

- ✓ HTML
- ✓ CHTML
- ✓ WML
- ✓ XML
- ✓ SyncML

All the preceding markup languages are derived from the *Standard Generalized Markup Language (SGML)*, a powerful, flexible, but complex language. Although they use some of the features and capabilities of SGML, each of the five languages listed above is designed to serve a more specific purpose than SGML, and does not require the full range of capabilities in that language. HTML, for example, is designed to describe the way information is displayed by a Web browser, and includes only features from SGML that are relevant to that task.

Each markup language has the same basic form, consisting of tags that provide information about the text they are associated with. The number, meaning, and interpretation of the tags differentiates one markup language from another.

HTML

HTML, the *HyperText Markup Language*, was designed to control how a Web browser displays text. Consequently, HTML tags specify such formatting and style attributes as Boldface, Italics, and Bulleted List. In addition, HTML tags can link text files together, allowing users to jump from one file (typically, a single Web page) to another by clicking the appropriate hyperlink on the screen.

Literally thousands of Web sites provide information about HTML. But for the final word on the subject, go to the source — *the World Wide Web Consortium's (W3C®)* HTML 4.01 Specification. This "recommendation," as the W3C calls it, completely defines the 4.01 version of the language. To read the specification, go to: `www.w3.org/TR/html401/`.

CHTML

Compact HTML (*CHTML*) is formed from a subset of the HTML 2.0, 3.2, and 4.0 specifications. As with HTML, most CHTML tags specify how text should appear on the screen. Most importantly, CHTML has been adopted and extended by NTT DoCoMo, and is the language of the "i-Mode" service. The extensions added by NTT DoCoMo add cellphone–specific capabilities, such as the ability to click a hyperlink and place a call.

To make CHTML function on cellphones and other mobile devices with limited processing power and small, variable-sized displays, several features of HTML are not included in CHTML:

- ✓ JPEG image display
- ✓ Tables
- ✓ Image maps
- ✓ Multiple fonts and character styles

Because CHTML is an extended subset of HTML, it is easier for Web developers to learn and develop pages for CHTML than WML (described in the next section).

A good source of detailed information on CHTML is a W3C Note entitled "Compact HTML for Small Information Appliances," which can be found at `www.w3.org/TR/1998/NOTE-compactHTML-19980209/#www1`. For CHTML information that is specific to i-Mode sites, try this link at DoCoMo Net: `www.nttdocomo.com/i/tagindex.html`.

WML

The *Wireless Markup Language* (*WML*) is a markup language based on XML. It defines the way content is displayed on devices using the Wireless Access Protocol (WAP). WML is designed to be used by small, narrowband wireless devices (the current generation of wireless devices). It fills the same role for WAP devices that CHTML fills for i-Mode devices.

WML tags specify four kinds of information:

- ✓ **Text presentation and layout**
- ✓ **Deck/card organization:** All information in WML is organized into cards and decks. Each card covers a specific user interaction (a menu selection or text entry field, for example). A group of cards is combined into a deck, with each deck roughly equivalent to one HTML page.
- ✓ **Inter-card navigation and linking:** WML allows users to navigate between cards and decks manually. (Users can also navigate under software control.)
- ✓ **String variables and state history:** WML supports runtime substitution of strings for variables, and tracks state changes within a session to allow for easier backward navigation.

In HTML, devices generally communicate directly with the HTML server. In WML, devices generally communicate with a server known as a *WAP Gateway*. The gateway handles communication with the WAP devices and translation of data from the HTML server.

WML is important because so many mobile phones support WAP.

You can find more information about WAP in Chapter 11. Go to the WAP.com Web site for the latest information on all things related to WAP, including WML (`www.wap.com/`).

XML

XML, the *eXtended Markup Language*, was created to allow the use of richly structured documents on the Web. Several markup languages relevant to wireless devices are based on XML, including SyncML and WML.

XML is another descendant of SGML, superficially similar to HTML but designed with a very different goal in mind. Whereas HTML exists to specify how text should be displayed on the screen, XML exists to identify different sections of text of a certain type. The interpretation of that type is left up to the interpreting program or device. With XML, it is possible to encode anything from e-mail messages to financial transactions to configuration parameters to spreadsheets or word processing documents in the form of tags and text. Think of XML as a set of rules for designing text formats that are easy for computers to create and interpret.

Another key feature of XML is extensibility. That is, XML is explicitly designed to support the creation of application-specific languages such as WML or SyncML. Extensibility is possible because — in contrast to all valid HTML tags, which are defined by the W3C — XML tags can be defined by anyone. To define a new language based on XML, you only need to define the new tags and how they should be interpreted. The new language may not be adopted by anyone other than the creator, but at least the ability to create the new markup language exists.

Clearly, XML is flexible and useful in a wide range of applications. However, one question remains: "Why choose XML or a language based on XML to describe my data?" The answer is implied in the question. Because XML uses tags to delimit data (in the form of text), and because new tags can be created as necessary, containing unlimited amounts of machine-readable information, XML documents can completely describe the data they contain. This eliminates the requirement for separate rules on how to interpret the data in a document, and greatly increases the potential power of search engines and data mining operations.

XML and its related specification are under the control of the W3C. The specification can be found at `www.w3.org/XML/`. In addition, two Web sites are dedicated exclusively to XML: XML.org (`www.xml.org`), which is run by the Organization for the Advancement of Structured Information Systems (OASIS) to minimize overlap in XML languages and XML language standard initiatives, and XML.com (`www.xml.com`), which is run by O'Reilly & Associates, Inc, with a mission to help individuals learn XML and use it to solve their information management and e-commerce problems.

SyncML

SyncML, the *Synchronization Markup Language*, is an XML-based language designed to be an international standard for wireless data synchronization. The goal of the organizations working on SyncML is to create a *Common Mobile Data Synchronization Protocol* to ensure that wireless devices can synchronize with networked data. The standard is designed to function on all major transport layers, including HTTP, WSP, and OBEX.

In essence, SyncML is a language for describing the process of synchronizing two or more devices over a network. SyncML does not define how an application uses the synchronized data, or even how to synchronize the data once an application has it. What it does do is enable devices and synchronization applications to communicate using a common language.

The SyncML protocol supports the following types of information:

- ✓ Naming and identification of records
- ✓ Commands to synchronize local and network data
- ✓ Identification and resolution of synchronization conflicts

SyncML is important because currently, each device, PIM, and synchronization application uses its own synchronization protocols, which limits each device or PIM to synchronize only with a limited number of other devices or PIMs. You can deal with this issue by creating specialized software for each application or device that must be synchronized, as Pumatech, Inc. does with its Intellisync software. But a much more efficient and practical approach is to have a common synchronization protocol that all devices and PIMs use.

Note that the worldwide adoption of a single synchronization protocol will not eliminate the need for synchronization software. Again, SyncML ensures that devices and PIMs can communicate with each other. What they do with the data once it has been communicated is determined by their synchronization software. SyncML just shifts the battleground from the number of different devices and PIMs each synchronization product supports to how well each synchronizes the data.

The official source of online SyncML information online is: www.syncml.org/.

Java

Java is a platform-independent, object-oriented programming language that is becoming increasingly important for wireless devices. While getting into the technical nitty-gritty of Java programming is outside the scope of this chapter, this section will provide some basics, along with the reason why Java is so important for the future of wireless devices.

From the beginning, Java was designed to work on many different hardware devices. It would be impractical to rewrite every program for many different types of hardware, so the designers of Java came up with a very clever solution.

Like most modern computer languages, Java is compiled. That is, the human-readable code that the programmer writes is converted into a machine-readable form. But whereas most compilers convert the human-readable code into a form that is directly executed by the computer, Java compilers create something called *Java bytecode*. Furthermore, Java bytecode is independent of the particular hardware it will run on. In other words, it is *platform-independent*.

When someone runs a Java program on a particular piece of hardware (a wireless phone, for example), the Java bytecode is run by an interpreter program that implements what is known as a *Java Virtual Machine* (*JVM*). The JVM interprets the Java bytecode into commands that the actual hardware can understand and execute. It is the JVM that talks directly to the hardware, so it is the JVM that must be written specifically for each type of hardware. Once a JVM is available for a particular type of hardware, any Java program can run on it.

Java will become increasingly important for wireless devices because of its platform-independence. Many different types of wireless devices are currently becoming available, and an even greater variety should start appearing in the next few years. The benefit of writing a Java program that, for example, keeps track of contacts and will work on any wireless device with a JVM is clear. Java also offers benefits such as built-in security and the ability to download the components needed to make a Java application run. Such benefits make Java a strong contender for the title of top programming language for wireless devices.

Luckily for us, information about Java is freely available online. Sun Microsystems ensures that all official information about the language is available on the Web at: `http://java.sun.com`. This site is rich in information, including goodies like white papers, full product documentation, and complete developer tools.

Wireless-specific information is available on the Java Wireless Developer Initiative page: `http://developer.java.sun.com/developer/products/wireless/`, while up-to-date information on the wireless devices that have a JVM can be found at JavaMobiles.com: `www.javamobiles.com/`.

OBEX

OBEX, short for OBject EXchange, is a set of high-level protocols that define the way data objects like vCard contacts can be exchanged using various technologies, including Bluetooth. OBEX support is built into a variety of devices and operating systems, including mobile phones from Siemens and Nokia, Microsoft Windows 2000 and Palm OS–compatible devices. The OBEX protocols are optimized to work over ad hoc wireless connections, and their use helps to reduce the snarl of incompatible standards that hinders the interconnection of wireless devices.

OBEX provides four primary functions: `connect`, `disconnect`, `get`, and `put`. `Connect` and `disconnect` initiate and terminate wireless communication sessions, while `get` and `put` retrieve or send data through the established communication session. Wherever possible, OBEX uses industry-standard data objects such as *vCard, vCalendar*, *vMessage,* and *vNote* to do its work.

You can find the latest version of the OBEX specification at the IrDA Web site (where it is also known as IrOBEX): `www.irda.org`. As of this writing, the specification was at version 1.2 and was available as a PDF file at this site: `www.irda.org/standards/pubs/IrOBEX12.pdf`.

Given the number of devices that currently support OBEX, OBEX support is a desirable characteristic of any wireless device, although you will probably have to dig into the technical specs to see if it is implemented on any particular device.

Summary

We now conclude your tour of the diverse device technologies that are on the brink of making a real impact on the wireless space. As you can see, the extreme — and usually conflicting — demands of wireless devices led to the development of extreme technologies.

Chapter 14

Wireless Security

In This Chapter

Unfortunately, we do not live in a world where everyone respects the rights of others. Security is an ever-present concern, particularly in the wireless world where people shop from their mobile phones, carry handheld devices containing megabytes of confidential company information, and connect to corporate networks by way of wireless LANs.

This chapter surveys the state of wireless device security, both for mobile devices, such as cellphones and PDAs, and for stationary devices, such as the workstations and access points of wireless LANs. The chapter is divided into two sections, one for each of these broad categories of wireless devices. The two sections contain information on the following topics:

- ✓ 1G device security issues
- ✓ WAP security issues
- ✓ Mobile IPv6 security issues
- ✓ Palm OS, Windows CE/Pocket PC, Symbian, and RIM device security issues
- ✓ Wireless LAN security issues
- ✓ Suggestions for securing a wireless LAN

Mobile Device Security

Security has been an issue for mobile wireless devices since the first cellphones became widely available. For example, after 1G phones were developed, unscrupulous people started listening in on other people's conversations, even "cloning" their phones to get free service. Since that time, a low-intensity war has been waged, with the good guys trying to secure the wireless networks and devices and the bad guys trying to snoop, steal services, or simply cause harm.

The following sections provide a short summary of the security situation for various generations of mobile wireless devices, along with some related technologies. The devices themselves — which are now at risk for virus attacks and other nastiness — are discussed, along with more mundane threats, such as the devices being stolen and having all the valuable data extracted.

1G Security Issues

First-generation wireless devices — for example, analog mobile phones — don't seem to have been designed with security in mind. Getting the first cellular networks up and running was a tremendous undertaking in and of itself, leaving little time or energy for other concerns.

It isn't difficult to eavesdrop on AMPS (Advanced Mobile Phone System) phone calls. All you need is a simple scanner, like that used to monitor police-band radio traffic. Since AMPS uses analog signals, it is hard to encrypt the signal, the way you could if the signal was digital.

One particularly nasty trick used by phone hackers is to monitor AMPS calls and decode the identifying information needed to "clone" a phone. By programming the right information into a phone, the bad guys can make it appear to the network as if their phone is someone else's. The growing popularity of digital cellphones (which can encrypt voice data), as well as GSM growth (GSM is more secure than most other phone networks), has helped blunt the prevalence of cellphone cloning. Also, the industry-wide use of new electronic authentication and digital encryption has drastically reduced phone cloning.

Thus, if you are concerned with the security of your wireless phone calls or dislike the idea of someone impersonating you (your phone, anyway) on a public phone system, consider using systems that support a newer generation of wireless technology.

Current Wireless Security Strategies

This section presents some successful strategies that help make data safe during transmission and also while residing on a wireless device. One set of steps is available to an end user of wireless devices, and a larger set of options exists for enterprise users and mobile wireless service providers.

END USER PRECAUTIONS

You can take steps to transmit sensitive wireless data over a CMDA network. CMDA is based on the spread-spectrum technology used by the United States military for secure communications. With CMDA, an eavesdropper cannot easily predict exactly what frequency that sensitive data will be transmitting on.

Beyond basic CMDA, as an end user, you can do the following:

- ✓ Inquire whether the network uses a service that employs an advanced form of "frequency hopping," which makes eavesdropping difficult.
- ✓ Employ encryption software, such as that made available from Certicom, or a certificate service, like that provided by VeriSign.
- ✓ Identify when data is on a device that is not secure. For example, if a sensitive document is currently on a Pocket PC, the document can be zipped with a password while not in use. Once the document is used, it can be removed from the less-than-secure device.
- ✓ Make sure that all who have legitimate access to sensitive data employ a hardware token verification of some type before access. Different types of token verification include smart cards or bioverification devices, such as a fingerprint scanner.
- ✓ Consider transmitting data over WAP only, utilizing a proprietary, beefed-up version of WAP's security layer, Wireless Transport Layer Security (WTLS).

ENTERPRISE USER GROUPS OR SERVICE PROVIDER PRECAUTIONS

As discussed in the preceding end-user section, professional group users and service providers should also utilize frequency hopping and CMDA networks, employ the strongest encryption available, and require that all those who have access to the most sensitive data use physical tokens to access it. Also, a company may want to transmit sensitive data over a form of WAP.

Solution providers such as Certicom Trustpoint provide wireless data encryption over server connections. Note that so-called triple-DES encryption is pretty much impenetrable, but you must still be aware of security holes, which allow points of unauthorized access. Most often, gateways are security holes in which data encryption is impossible to guarantee. A company may therefore want to set up its own gateway and provide only private access to that data, develop a fail-safe relationship with the owners of the gateway, or deploy the most sensitive data wirelessly on a VPN network.

Following are some other actions that you may want to take to ensure your enterprise's wireless strategies are secure:

- ✓ Investigate security options that use asymmetrical keys for encryption/decryption, such as PKE.
- ✓ Make sure all transaction partners use strong encryption as well. If customers access your services wirelessly to carry out transactions, look for security holes in every stage of the transaction process, including middleware.
- ✓ Utilize strong verification measures at both ends of a transaction. Have customers, as well as employees, sign on with a password.
- ✓ When developing an application for wireless transactions, consider using Java, which is considered very secure.
- ✓ When writing code for a transaction process, make sure that details are saved in a readily available database. Include all knowable details about that transaction, including device ID, time and date, user ID, transaction amount, and so on.

GSM Security Issues

GSM signals are encrypted, so snoops would need to crack the code even if they do manage to eavesdrop on a call. The security password is stored on the SIM, not transmitted over the air. To clone a GSM account, you need to tamper with the tamper-proof SIM, which destroys itself after too many wrong passwords are entered.

128-Bit Encryption and International Export

The United States views the ability to strongly encrypt data as a possible security threat and would like to prevent 128-bit encryption technology from falling into unfriendly hands. If your company exports products that provide strong encryption, expect visits from men in little black hats asking lots of questions. Government agencies are not always aware how accessible encryption creation technology is for the asking on the Web. They expect good corporate citizens to prevent its proliferation via their internationally exported products.

WAP Security Issues

WAP devices use the WTLS for security. While this works fine, it is not the standard security protocol used by most businesses and the financial industry. As a result, WAP creates a potential security problem when you conduct online financial transactions (e-commerce).

Somewhere along the line between the WAP device and the rest of the economy, the WTLS-secured data must get converted to SSL-secured data. And during that process, for at least an instant, the data exists on some intermediate computer in a decrypted state. This normally occurs at a network operator's computer acting as the WAP gateway. Industry experts are divided on whether or not this conversion is actually a significant security problem, but at least theoretically, the data traveling between a WAP-capable device and an e-commerce site or corporate network is exposed to a third party as clear text.

Mobile IPv6 Security Issues

Some debate exists over the security of the proposed *Mobile IPv6* (*MIPv6*) protocol. MIPv6 would allow wireless devices to remain authenticated with the network, even as they physically move through it. Currently, devices (using IPv4) remain authenticated through their home IP addresses. As they physically move through the network, they must continually get new local IP addresses. But to remain authenticated, all communication to and from the network must travel through the original home IP address.

The plan for MIPv6 is to eliminate the need to relay all messages through the home address, while still remaining authenticated on the new local IP addresses. However, *IPsec*, the security protocol used by MIPv6, requires a global *public key infrastructure* (*PKI*) to do its work, and that infrastructure does not yet exist. Given that the creation and deployment of a PKI is a political as well as a technological issue, it is unclear when such an infrastructure will come into existence, if ever.

The solution to this problem is unclear right now, with industry experts and organizations such as the Internet Engineering Task Force working on the problem. Until PKI usage is resolved, the overall security of MIPv6 will remain unknown.

Wireless PDA Security Issues

Because wireless PDAs have essentially become small, general-purpose computers, they are vulnerable — like other general-purpose computers — to security problems, as well as to attacks by viruses, worms, and other nasty software (sometimes grouped under the general heading of *malware*). There have been few attacks on wireless PDAs to date, but vendors have started to address the problem for the small number of malware programs currently afflicting wireless PDAs.

Two types of security software are appearing for wireless PDAs: file encryption software and anti-virus software. The file encryption software protects files stored on the device by encrypting them and requiring the user to enter a password before decrypting the file. The anti-virus software detects viruses and other malware and takes appropriate action.

PALM OS DEVICES

Anti-virus and encryption software is available for the Palm OS from various vendors.

ANTI-VIRUS SOFTWARE McAfee provides a range of anti-virus programs for Palm OS devices, protecting the Palm OS device from malware that has been targeted against it, as well as protecting the desktop PC against PC viruses attached to files on the device. F-Secure also provides anti-virus software. In both cases, there is software resident on the device, as well as on the PC the device cradle is connected to.

ENCRYPTION SOFTWARE F-Secure and Certicom both provide file encryption software that resides on the device, protecting files the user designates, and requiring the user to enter the appropriate password or phrase before decrypting those files.

The McAfee Web site is `www.mcafee.com`. The F-Secure Web site is `www.f-secure.com`, and the Certicom web site is `www.certicom.com`.

RIM PAGERS

Security problems involving RIM BlackBerry pagers are apparently uncommon. This may be due in part to the fact that RIM included triple DES encryption technology in the Enterprise Edition of the device. It may also be due in part to the fact that the number of RIM pagers in use is still far lower than the number of Palm OS devices or Pocket PC (Windows CE) devices, making it a less inviting target for the bad guys.

SYMBIAN DEVICES

Anti-virus software support is available from McAfee for Symbian devices. At this writing, no third-party encryption programs are available. Given Symbian's exit from the consumer handheld market, it is unclear whether anyone will develop such applications.

WINDOWS CE AND POCKET PC DEVICES

Anti-virus and encryption software is available for Pocket PC and Windows CE devices from various vendors. However, the offerings here are more variable than they are for Palm OS devices.

ANTI-VIRUS SOFTWARE McAfee provides separate anti-virus software for Windows CE and Pocket PC devices. Neither of these products includes a component that resides on the device. While this frees the device memory for other uses, it means the device is only scanned for viruses when it does a wired synchronization with the PC. On the other hand, the anti-virus solution from F-Secure works for both Windows CE and Pocket PC devices and includes an anti-virus component that resides on the device and can be used to scan on command. This capability is especially useful for a wireless device, where the user can download files wirelessly. By scanning files as soon as they arrive on the device, users can give themselves a bit of extra security.

ENCRYPTION SOFTWARE F-Secure and Microsoft both provide encryption software that resides on the device, protecting files the user designates and requiring the user to enter the appropriate passphrase before decrypting those files. The Microsoft solution is free and may already be included on your device.

The Pocket PC section of the Microsoft Web site can be found at `www.microsoft.com/mobile/pocketpc/`.

Wireless LAN Security

The two security priorities for any network are the same: access control and privacy. In its most basic form, *access control* ensures that only authorized users can log on to the network. A network will often have levels of access, with administrators granted greater rights than regular users. In any case, the goal is the same — to prevent unauthorized users from connecting to all or part of the network.

The issue of privacy is slightly different. *Privacy* ensures that data traveling across the network can be read only by the intended recipient. In terms of networks, privacy means data encryption.

Providing security for wireless LANs is significantly more difficult than providing security for wired LANs. With wired LANs, ensuring access control and privacy is mostly a matter of physically securing the network. If an intruder can't physically make contact with the network, he can't violate its security. Although wireless LANs present the same security concerns as wired LANs, the wireless LAN has unique issues of its own. Wireless LAN data is broadcast omnidirectionally — rather than traveling along shielded cables within the building. The radio signals carrying this data can travel through windows, walls, and floors, right out of the building and into the receivers of any nearby snoops. The relatively transparent nature of data transmission is a problem fairly unique to WLAN systems, one that has garnered much media attention recently.

In this chapter, we assume that you know how to provide security for your wired network. The issues discussed here are specific to the security of the wireless portion of your network.

Security for an 802.11b Network

All the excitement over wireless LANs today concerns versions of the IEEE 802.11 standard. Most popular, and therefore most important where security is concerned, is the 802.11b standard. 802.11b provides mechanisms for both access control and privacy, although it now seems that none of these mechanisms are adequate to the job of truly securing a wireless LAN.

Understanding War Driving

Many wireless networks fail to implement even the most basic security measures, giving rise to the phenomenon of *war driving*, the act of cruising the streets of a town looking for vulnerable wireless networks. If you are looking for unprotected networks, all you need is a laptop with an 802.11b wireless network card and a custom antenna located on the roof of your car. To take things to the next level, you could include your favorite 802.11b hacking software on the laptop and try to break into secured networks.

According to computer security consultant Peter Shipley, hundreds, perhaps thousands, of insecure 802.11b networks exist in the region from San Francisco to Silicon Valley alone. No one knows the exact number of people cruising around towns, looking for wireless networks to snoop on, but there are certainly people engaging in this type of activity. If you do valuable work while connected to a wireless network, make sure you follow the security steps outlined in this chapter.

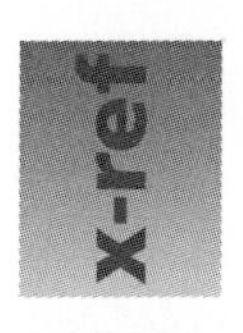

For details of the 802.11b standard, see the 802.11 sections in Chapters 5 and 11.

SERVICE SET IDENTIFIERS

The access control component of 802.11b is called the *Service Set Identifier* (*SSID*). The SSID provides a common network name for the wireless network interface cards and other devices in a wireless LAN. Any device that does not transmit the proper SSID cannot connect to the wireless LAN.

Unfortunately, SSIDs are a weak form of access control. In general, it is easy to determine what the SSID for a wireless network is — most of the time, it is broadcast to anyone listening by the wireless access point.

WIRED EQUIVALENT PRIVACY

The privacy provisions of 802.11b are stronger than the SSID access control but not strong enough for many applications. *Wired Equivalent Privacy* (*WEP*) is an optional security feature of 802.11b. Since support of the 40-bit version of WEP is a requirement imposed by the *Wireless Ethernet Compatability Alliance* (*WECA*) for a device to receive *Wi-Fi* certification, virtually all 802.11b devices support 40-bit WEP.

WEP is used for both access control and privacy. Without the correct WEP key, a device cannot connect to the network. Even with a valid WEP key, only data encrypted with that key can be decrypted by a device connected to the network. At least, this is what WEP is designed to accomplish.

WEP is a symmetrical key encryption system that uses the same algorithm and key to encrypt and decrypt the data stream. Therefore, to function on the network, a device needs a valid WEP key. The 802.11 standard defines two ways to distribute WEP keys to devices that want to connect to the network.

To distribute WEP keys, up to four default WEP keys can be shared among all devices connected to the wireless LAN, including access points and NICs. The advantages of this approach are clear. It is easy to distribute and maintain the WEP keys, and it is also easy to add new devices to the network. Unfortunately, the easier it is to distribute keys and to add additional devices to the wireless LAN, the more likely it is that the entire LAN will be compromised. Once the WEP keys are known by an outsider, it is necessary to change all the keys on all the devices to resecure the network.

Another more secure method for distributing WEP keys is through a mapping table. With this approach, a unique key is assigned to each device, and each device is identified by its *Media Access Control* (*MAC*) address. An intruder needs both the MAC address and the associated WEP key to gain entry to the network. However, a patient hacker can obtain the MAC address by monitoring the wireless access point control and data channels. A more impatient or daring hacker can simply steal a device, gaining both the MAC address and the WEP key simultaneously.

Authenticating a Device on an 802.11b Network

Before a device can connect to an 802.11b LAN, the LAN must authenticate that device. The 802.11b standards support two types of authentication: *open* and *shared-key*. A third type of authentication, based on the device *MAC address,* is available from some 802.11b suppliers.

OPEN AUTHENTICATION

By default, 802.11b networks use open authentication, which allows any 802.11b device to connect to the network, even without a valid WEP key. Open authentication is done without the use of encryption. Any 802.11b device can associate itself with an access point using open authentication.

SHARED-KEY AUTHENTICATION

In shared-key authentication, a wireless LAN access point sends the device a challenge text packet. The device must use its WEP key to encrypt the challenge text packet and then send it back to the access point. If the device encrypted the challenge text packet with the proper WEP key, the device is granted access to the network.

MAC ADDRESS AUTHENTICATION

MAC address authentication depends on the fact that the LAN maintains a table of MAC addresses for devices that are allowed access to the LAN. A device is only granted access to the network if it has an approved MAC address. MAC address authentication can be used with or without a valid WEP key.

Vulnerabilities of 802.11b Networks

While the 802.11b standard provides some access control and privacy protection, wireless LANs based on the standard have several security vulnerabilities. These include the following:

- ✓ **Stolen devices:** Wireless devices, particularly mobile devices such as laptops and PDAs, are relatively vulnerable to theft. Unless device log on is password protected, the thief will have access to the LAN. Such a theft will be difficult or impossible for the network or the administrator to detect and will likely go unnoticed until the legitimate owner of the device reports the theft.

- **User impersonation:** Multi-user devices present real problems for LANs that use the MAC address as part of the security scheme. If three people use such a device, all three will have the same level of access to the network and to each other's data. Without additional security measures, it is impossible to determine which authorized user is using the device at any given time or even if the user is authorized.
- **Roaming users:** This problem is related to the problem of user impersonation. What happens to the security scheme if a user wants to log on to the LAN using a different device? If all the security measures are based on device characteristics like the MAC address, the network will not be able to tell which user is connected through a particular device. If Joe uses Sally's device to connect to the wireless LAN, the LAN will give Sally's network rights to Joe, as well as access to her data, not his.
- **Rogue access points:** On an 802.11b LAN, access points authenticate devices; devices cannot authenticate access points. By placing an unauthorized access point on a wireless LAN, a hacker could gain access to devices connected to that network, using them to launch denial-of-service attacks.
- **WEP encryption failure:** Researchers from Cisco and the Atlanta-based Internet Security Systems have shown that they can defeat WEP encryption through one technique or another. Indeed, a product was released in late 2001 called AirSnort that allows Linux users with a wireless network card to snatch passwords and other data from an air link.

Defeating WEP Encryption

As of this writing, numerous reports have appeared describing ways to defeat the WEP encryption scheme used by 802.11b. These stories do not involve lost or stolen devices or WEP keys. Instead, they describe methods of defeating the encryption system itself. The next few paragraphs describe two ways in which the WEP system can be defeated.

One approach relies on the very nature of the WEP and the way it must be implemented in a wireless environment. WEP uses an encryption algorithm known as *RC4*. RC4 is a *stream cipher*, which operates on a continuous stream of data. RC4 uses an *initialization vector* (*IV*) to start the encrypting process. This cipher is very effective when the data stream is reliable. In a wireless environment, data packets are frequently lost, which disrupts the ciphering process.

To ensure that the decrypting of data can continue, WEP transmits a new IV with every data frame. Since WEP uses a 24-bit IV, it does not take very long (usually no more than a few hours) for a typical 802.11b LAN to reuse an IV.

Reusing an IV isn't necessarily a problem. It is, however, a real problem on LANs that use static WEP keys or that change the WEP keys infrequently. To understand why, you need to understand how the IV is used in WEP.

Continued

Defeating WEP Encryption *(Continued)*

To encrypt data using RC4, the transmitter performs the following steps:

1. Appends a new IV to the WEP key to create an RC4 *keyschedule*.
2. Generates a *keystream* that is the same length as the payload (the data to be transmitted, plus a *CRC*).
3. XORs the keystream with the payload to generate the *ciphertext*.
4. Transmits the IV (unencrypted) and the ciphertext.

To decrypt the ciphertext, the receiver performs the following steps:

1. Appends the newly received IV to the WEP key to create an RC4 *keyschedule*.
2. Generates a *keystream* that is the same length as the payload (the data to be transmitted, plus a *CRC*).
3. XORs the keystream with the ciphertext to decrypt the payload.

Since the system generates new IVs continuously, and since those IVs are transmitted unencrypted, a hacker can monitor the network traffic and record multiple data frames that use the same IV. Assuming that the WEP keys do not get changed frequently, the hacker can gather lots of data when the WEP key remains the same and the IV is known. With this information, the hacker can perform statistical analysis. 802.11b frames contain several IP packets where the data is known. This allows the hacker to generate a partial keystream every time the same IV appears. Eventually, the hacker can crack the entire code, and will be able to decrypt all frames until the WEP key is changed. Reportedly, this process is simple enough for a hacker with a standard PC and a good-sized hard disk to successfully crack an 802.11b LAN where the WEP keys do not get changed frequently.

Even worse, as this chapter was being written, there were reports of a new method for attacking WEP encryption. With this approach — known as the Fluhrer, Mantin, and Shamir attack — a graduate student at Rice University, working with a researcher at AT&T Labs, was able to use a wireless LAN card and two hours of programming time to determine a 128-bit WEP key used on a production network. 128-bit WEP keys are part of a stronger version of WEP security, slated to become available toward the end of 2001. Furthermore, the team accomplished this task passively — that is, by simply monitoring network traffic without transmitting to the wireless network at all. This means there is no way to detect that network security has been broken.

You can read a copy of the initial report detailing this attack at the Rice University Web site: `www.cs.rice.edu/~astubble/wep/wep_attack.html`.

Securing an 802.11 Network

Now that you've heard the security horror stories around 802.11 wireless networks, you may be thinking that the risks of going wireless outweigh the benefits. While it would seem imprudent to allow wireless access to all corporate network data, you could choose a middle ground. A possible course of action is to take all the steps you reasonably can to secure wireless access while keeping your most crucial data on the wired portions of your network.

One of our favorite places to go for up-to-date information on wireless (and wired) networking is Practicallynetworked.com. This site has a useful section on securing a wireless network, with a large list of security tips, plus links to many other related sites. The URL is `www.practicallynetworked.com/support/wireless_secure.htm`.

Here are some steps you should take to minimize the risk to your network security:

✓ **Keep your wireless access points outside the main network's firewall.** In effect, this treats your wireless network as an insecure external network.

✓ **Enable WEP.** It may seem silly to do this after reading about the ways that a hacker can break WEP, but it's like locking your front door at night — you won't slow down the determined intruder, but you may deter casual snoops.

✓ **Change WEP keys frequently.** This can be a lot of work, but changing the keys will at least force determined intruders to break into the network again.

✓ **Enable any user authentication features available on your wireless network.** By the time you read this chapter, 802.11b networks that support the proposed 802.1X Port Based Network Access Control standard may be available. 802.1X adds central management and user authentication to 802.11b networks. This standard or any other system that requires a network log in (entry of user name and password) to gain access to the network will improve security against active intrusions into your network.

Summary

As this chapter has made abundantly clear, you should be aware of some real security issues when using wireless devices, especially since IEEE 802.11b LANs are proving to be vulnerable to a variety of attacks. Although you can take steps to make your wireless LAN more secure, you should assume that a determined hacker will be able to break in.

Chapter 15

Introduction to the Wireless Web

In This Chapter

In this chapter, we discuss requirements for creating a wireless Web site, including the differences between how wired and wireless Web sites appear and function. You learn various approaches towards developing a wireless Web presence, the different technologies involved, and how to apply them. This chapter discusses the following:

- ✓ The requirement for standards
- ✓ How wireless and fixed-wire Web sites differ
- ✓ Wireless markup language approaches
- ✓ The Wireless Application Protocol
- ✓ The Wireless Markup Language

The Standardized Wired World

Every wireless communication device manufacturer and service provider would like to make the wireless Web a pervasive reality. If consumers started to equate their cellphone or PDA with Web access, the wireless industry may well be able to replicate — or even supersede — the stunning success of the PC revolution. However, a set of common standards for Web-page viewing and data transmission is needed before this can be accomplished. Many technologies currently provide a type of Web experience to handheld users, but the uniformity of basic Web access technology enjoyed by the PC world is not yet a reality for wireless.

PC-based Web technology became ubiquitous partly because everyone backed one horse: the TCP/IP stack and related protocols. Almost every desktop or laptop computer that accesses the Web evokes the same set of network standards, which means that, from machine to machine, our Web-viewing experience is remarkably similar. Uniformity has meant that Web content providers have a clear idea as to how a "Web site" should look and feel and can somewhat predict what a user will see when a site is accessed. Uniformity has also meant that PC manufacturers can promise a consistent Web experience as they hawk their wares to the public.

Uniformity and Realism in the Wireless World

In the wireless Web world, we've a bit of hashing out to do. Almost every cellphone model uses a slightly different approach to displaying Web content. Service providers implement a number of access solutions, each involving use of legacy network equipment. It's quite risky for a provider to replace existing equipment and technology and embrace a standard that may or may not take root with device manufacturers or customers.

Fortunately, the industry at large has stopped over-hyping the experience of accessing the Internet from a small wireless device. Promises that you can "surf the Web from your cellphone" have given way to more realistic assertions that you can "access the information you need." Industries have developed technologies that seriously address the intrinsic limits of the wireless Web. As expectations are lowered from mania to healthy enthusiasm, it appears you can do quite a bit online with your cellphone or PDA.

This chapter discusses the challenges of applying familiar Web technologies to the world of wireless and how those challenges have been met. It will familiarize you with the various spins on HTML and XML that are now being applied to wireless, as well as the new protocols and standards. The Wireless Application Protocol (WAP) and the basics of creating a page with this pervasive technology will be covered.

Getting to the Wireless Web

Before addressing the specific challenges that wireless Web-builders face, we must ask these fundamental questions:

- ✓ Does creating a wireless Web mean transforming existing sites to a miniature format or creating new content from scratch? With more than a billion "registered" Web sites in existence, you can see why it is tempting to envision a magic button that will "insta-convert" all that hard work.
- ✓ Given the limitations of the wireless Web, can content providers create sites that are as enriching and popular as their big-screen counterparts? Can Web designers stretch (or rather shrink) their imagination to embrace this new medium?
- ✓ Will the world at large catch on to the value of Internet via 160 × 160 screen? Can Web surfers renounce the concept of casual browsing and clicking and embrace a more intentional, "information now" approach?

The Wireless Web Challenge

Whether you are a Web content creator, service provider, or device manufacturer, the wireless Web poses the following challenges:

- **Low data rates:** Currently, cellphones provide 9.6 or 14.4 Kbps data speeds. Some wireless GPRS overlay solutions offer 19 Kbps. However, even the most advanced technologies on the burner will bump GPRS speed only up to 56 Kbps. Only those users who access the Internet via a wireless LAN will be downloading at DSL or cable speeds.
- **Small screen size:** Most cellphones can display only a dozen or so lines of text. The small screen size of cellphones, PDAs, and Pocket PCs severely limits how much information a Web page can convey. The spoils will not go to the swift but to the brief and succinct.
- **Irregular screen size:** Web designers often take the ratio and dimensions of a desktop computer monitor for granted. Regardless of resolution or display quality, every PC monitor has a 4:3 landscape aspect ratio. Web page content is designed to make good use of this orientation, giving the viewer a comfortable "eye-full" of text and images before prompting a click or a scroll. Compare that uniformity to what you find in handheld wireless devices.
  - Mobile phones lean towards a 3:2 aspect ratio.
  - Palm displays are often square.
  - Pocket PCs are usually 3:4 or something similar.
  - "Checkbook-style" handheld PCs are 2:1.
- **Use of color:** While most PDAs and Pocket PCs on the market today have at least 8-bit color displays, not all do. Web designers should take a cue from the fixed network world — where every attempt is made to design pages to the lowest common denominator viewing equipment — and avoid color graphics if at all possible.

Migrating Towards the Future

A migration needs to occur — from the norms of fixed-wire network Web viewing to a wireless Brave New World. To begin, Web content for wireless handhelds must be optimized for low data rates and small, irregular screen sizes. What else must change?

ACCOMMODATE FLAWED CONNECTIVITY

In the wireless world, dropped calls and lost connections are the norm. That means wireless Web sites should not have data collection that takes you back to square one if you get cut off. Sessions must be resumable. Presumably, the ability to restore a session should be taken care of in the application layer of a wireless protocol. Also, due to unpredictable connection loss, the user must be able to selectively resend message portions. You should not have to resend an entire lengthy message because of a disconnect. These issues are more than an irritant, since many wireless Web users will be dipping into their monthly cellphone minutes while online.

DON'T OVERLOAD THE PHONE

A client-server approach should be taken. Only a simple microbrowser, for example, should be incorporated into the mobile phone. The microbrowser-based services and applications should reside temporarily on servers, not straining the resources of the phone. When you want to add new features and services, avoid adding plug-ins or software-resident programs to the phone;

instead, enrich the functionality of the network. (However, manufacturers must then agree on the best microbrowser to deploy all around.)

INCREASE LATENCY PERIOD BEFORE RETRANSMITTING DATA

Because information exchanged over the wireless Web must go through a large number of gateways and other network elements, allowable latency must be increased. In the standard Web protocol, the TCP layer uses a relatively short retransmission timer value — a shorter latency period — before retransmitting data. If that same latency period were applied to a wireless Web transaction, the network layer would resend lots of information unnecessarily. In the wireless world, a new approach must be taken to determine when a network should resend "lost" information.

MOVE TOWARDS PACKET-SWITCHING

The wireless Circuit Switching Data (CSD) channel approach currently used by many GSM, CDMA, and TDMA services results in charges-per-minute for wireless Web users. Customers cannot afford this for long. When charged for time online, customers will regard wireless Web access as a pricey luxury. Service providers must be able to provide Web customers with a wireless packet-switched (IP layer) connection of some sort over CDPD, GPRS, or cdma2000.

The Quest for a Developer's Language

The wireless Web needs a language for developers, probably a markup language of some sort. Protocols are also required. The following sections discuss how these particular challenges are being met.

Is HTML Workable in the Wireless World?

Many question the fate of HTML in the wireless Web world. Does HTML have to go? Some say yes. For example, HTML is an interpreted (not compiled) language. It is transmitted as ASCII code straight "as is" to the browser, where it can be formatted. For example, the TCP/IP protocol will transmit a text string of 2,000 space characters. This transmission alone would account for half the size of a typical wireless Web page. A binary-encoded language is preferable for wireless Web design, reducing the amount of sheer data transmitted between handset and server. In the WAP-based language WML, for example, you can use all the comments you like while coding, because comments are not transmitted with the page — only encoded data. When designing your WML (Wireless Markup Language, the language of WAP) page, your comments will sit on the server and not be transmitted to the user.

In addition, HTML has been "accessorized" over time. Animations, colored fonts, cascading style sheets, and even most graphics are not going to work in a microbrowser environment, which the wireless Web requires. Even HTML "basics" such as frames, image maps, and tables are not supported enough, or not sufficiently supported enough, to warrant use.

Good Conscientious Coding

Some wireless "misses" can be avoided simply by considerate Web design practice. Consider, for example, when you use the venerable `password` input attribute found in HTML forms, like this:

```
<input type="password" name="shopper">
```

With this attribute in place, asterisks are displayed when the user types their password. However, from the point of view of a cellphone user, it is not helpful to eliminate visual confirmation of letters entered. Each button on the phone is used to input three letters. Without a visual cue, you won't be sure which letter you just entered. Boom. Wrong password. Start over. At 25 cents a minute? Not likely.

C-HTML

One new approach to wireless Web page development has been the implementation of an HTML subset called *C-HTML*, or *Compact HTML*. Some HTML purists would call C-HTML "HTML as it should be." Frames are a no-no, as are colored and stylized fonts. C-HTML is really a simple commitment on the part of the Web designer to avoid using HTML features that are going to be lost on a wireless audience anyway. Also, a C-HTML page should be no longer than four kilobytes. Beyond these limitations, creating a C-HTML Web page is really a matter of including the following code line at the top of your page (which indicates the type of page to the browser):

```
<!DOCTYPE HTML PUBLIC "-//W3C//DTD HTML 3.2//EN">
```

C-HTML has been adopted by NTT DoCoMo, the Japanese megaprovider. This company alone provides Internet service for millions of i-mode users who have enjoyed Web access via cellphone for some time now. C-HTML is not widely used in North America or Europe. However, the fact that C-HTML is the mobile data standard for Japan's i-mode phones means that C-HTML is a success. NTT DoCoMo was the first mobile operator to roll out packet-switching networks, enabling "always-on" access for i-mode customers. This is a tribute of sorts to C-HTML, since, at a minimum, the standard did not interfere with the packet-switching upgrade. In wireless, when existing standards work with new technology rollouts, everybody has a reason to smile. It doesn't always happen that way.

Web Clipping

We discuss Web clipping in Chapter 2, but this section covers it from the point of view of network functionality.

The Web Clipping Application (WCA) is a proprietary system developed for the Palm VII organizer. It runs as a subset of HTML, and a Web designer can create Web clipping pages by applying a few simple HTML tricks — and unlearning others. Palm VII PDAs use Bell South's Mobitex network, which somewhat limits the availability of Web clipping pages. However, other devices can view them now as well.

The Palm VII cannot interpret HTML. Rather, a proprietary program called Palm Query Application (PQA) is read directly by the Palm unit. When a site is visited, the PQA tells the Palm which page segments need to be downloaded and which have not changed since the last visit (and can thus be viewed offline).

PDAs that can only view Web clipping pages have a very limited Web experience. However, the PQA can download an entire site at once, rather than a page at a time. Once a site is downloaded, the viewer can enjoy instant switching from one page to another. And since the Palm knows to download only new, updated content, site access with the Palm VII can be satisfyingly peppy.

The Palm Web site provides a wealth of literature and tutorials for designing Web clipping sites. As with C-HTML, designers should avoid named fonts, style sheets, image maps, frames, and other extended HTML attributes. In general, keep the following points in mind when developing Web clipping pages:

- ✓ Web clipping pages are usually very small, avoiding paragraphs of text and images and instead offering the user a mode of interaction to obtain a result (for example, "Find a City," "Calculate Subtotal," and "Current PG-Rated Movies").
- ✓ When creating links, developers must know which instances to link to the PQA, or to the page itself, for a Web clipping page.
- ✓ You must include the metatag `<meta name="palmcomputingplatform" content="true">` at the top of your page.
- ✓ When you create a Web clipping page, you create the PQA simultaneously. You can then offer the Web clipping page to the world at large — it joins the selection of pre-set pages readily available to all users, just like the Barnes & Noble or Yahoo sites. Viewers can link to and download your page onto their Web clipping-enabled PDAs, viewing it offline, automatically updating its content when they HotSync, and so on.

Handheld Device Markup Language

The Handheld Device Markup Language (HDML) was the first mobile wireless-specific markup language. Developed in 1997 by Unwired Planet (which went on to become Phone.com and then OpenWave), it was also the first major industry attempt to build a wireless-aware site development language from the ground up. HDML introduced two radical concepts:

- ✓ Writing code for a cellphone's programmable "soft keys" — the unlabeled keys you often see at the bottom right and left of a newer cellphone (unlabeled because application developers can include instructions on their use in their program). The most common uses for these soft keys are "Back," "OK," and "Next."
- ✓ The card and deck site structure that HDML introduced to replace Web pages with single-action cards.

CARDS AND DECKS

A *card* is a prompt or display of information that fits comfortably on the smallest cellphone screen. The medium forces developers to think small and to think action. In HDML, you can program a card to do one of three things:

- ✓ **Display information:** The viewer can read the information and press Next or Back — that's all.
- ✓ **Prompt data entry:** The viewer can be prompted to type a bit of data into a field, such as a phone number, credit card number, or password.
- ✓ **Prompt a choice:** The viewer can be prompted to indicate a preference, such as a choice from a list of restaurant types. The next card can prompt the viewer to select a specific restaurant. These cards can also be thought of as link cards; when the viewer clicks a preference, they are really clicking a link.

Every site you create in HDML is based on cards, and each card can perform only one of the three preceding tasks. A set of cards is called a *deck* and is roughly analogous to an HTML-based "Web site," since you'll create a whole menu system — linked information, branching product description, and so on — with connected cards.

The only major difference between an HDML deck and a Web site is that the deck is a single file — not the cards themselves. When you create an HDML deck, you create all the cards at once. Each card is short, so the task is not as burdensome as it seems.

THE HDTP PROTOCOL

Besides developing HDML, Unwired Planet developed the *Handheld Device Transport Protocol (HDTP)*, a lightweight protocol for carrying out client/server transactions.

Unwired Planet also created a microbrowser called the UP.Browser. The UP.Browser has gone through three iterations since its creation, and today it is one of the most standard microbrowsers in the U.S. In fact, when some service providers say they are WAP-enabled, what they really mean is that you can view HDML pages on the OpenWave UP.Browser 3, with a few security protocols thrown in, which leads us to the story of WAP.

The Wireless Application Protocol

The future of HDML once looked very mixed, at best. Although AT&T Wireless deployed HDML/HDTP enthusiastically on a number of sites in 1996, HDML in the mid-to-late 1990s was all dressed up with nowhere to go. No major manufacturers were jumping on an HDML/HDTP bandwagon.

In 1997, the U.S. service provider Omnipoint offered a tender to anyone who could come up with a workable mobile Web service platform. The big catch was that it had to be nonproprietary — no Microsoft-ian locking the wireless world into one device or platform. What was required was a broad but practical standard that could be deployed on lots of devices. Four major cellphone vendors stepped forward to suggest their own solutions, while Unwired Planet submitted HDML.

Omnipoint cleared its throat and reiterated "no proprietary solutions." Everyone would have to get together and hash out a standard. The result was the Wireless Application Protocol (WAP) forum. A year after its founding, the first version of WAP was launched (see Figure 15-1).

Figure 15-1: A WAP page displayed on a cellular phone

WHAT IS WAP?

WAP is unique because it is not just a partial solution. WAP is a completely new stack of protocols. As a stack of protocols, the standards address wireless challenges that have to be dealt with in more than one network layer. Earlier in this chapter, we indicated that a true mobile wireless system needs sessions to continue after disconnect and also needs to take full advantage of packet switching. WAP does just that — it implements a stack. It embraces the microbrowser-gateway concept. The Wireless Application Protocol stack includes the following:

- ✓ Wireless Markup Language.
- ✓ WBMP (Wireless Bitmap) image format.
- ✓ Wireless Telephony Application.
- ✓ WMLScript.
- ✓ WAP gateway — microbrowser architecture.

Because WAP is designed across layers, it is easy to extend. Developers can create new content for one layer, leave the rest of the stack intact, and still "deploy WAP." The WAP protocol stack is divided into five different layers. The WAP protocol includes specifications designed for each network layer, allowing the creation of WAP software that meets standards at each level.

- ✓ **Transport layer:** Wireless Datagram Protocol (WDP); also can use UDP/IP.
- ✓ **Security layer:** Wireless Transport Layer Security (WTLS).
- ✓ **Transaction layer:** Wireless Transaction Protocol (WTP); considered part of the session layer, technically.
- ✓ **Session layer:** Wireless Session Protocol (WSP).
- ✓ **Application layer:** Wireless Application Environment (WAE).

The Wireless Application Environment supports three unique applications:

- ✓ **WML:** The Wireless Markup Language.

✓ WMLScript: A Java-based scripting language for wireless.

✓ WTA: The Wireless Telephony Application. Using the WTA, WAP can access phone services (for example, use addresses from a contact list).

WML — AN XML VARIANT

The Wireless Markup Language (WML) is an XML-based variant of HDML. Since WML pages will be deployed on all kinds of devices (mobile phones, PDAs, Pocket PCs, even televisions), a language that lets the developer create their own tags is a must. Because WML is a HDML subset of XML, developers can "write once, deploy everywhere" after designing a few flexible tags.

HOW WML WORKS

WML can handle events, carry through actions, and such. When you write in WML, you'll be using familiar scripting and XML objects using operators, variables, functions, and function calls. Some of the unique tags frequently used in WML include the following:

✓ `<card> . . .</card>`: Cards make up the contents of a deck. There are many types of cards.

✓ `<do> . . .</do>`: The `do` tag performs actions. Actions affect cards or decks.

✓ `<fieldset> . . .</fieldset>`: Use the `fieldset` tag to group several input or text elements within a single card. Facilitates quick card manipulation.

✓ `<go> . . .</go>`: The `go` tag navigates to a URL (for example, `go href="newcard.wml"`)

As with C-HTML, a header is required at the top of the page to inform the browser of the page type. The WML `DOCTYPE` identifier is:

```
<?xml version="1.0"?>
<!DOCTYPE wml PUBLIC "-//WAPFORUM//DTD WML 1.1//EN"
"http://www.WAPforum.org/DTD/wml_1.1.xml">
```

Take a quick look at a bit of WML code. You'll see that it is quite simple:

```
<?xml version="1.0"?>
<!DOCTYPE wml PUBLIC "-//WAPFORUM//DTD WML 1.1//EN"
"http://www.WAPforum.org/DTD/wml_1.1.xml">
<wml>

<card id="mycard" title="mycard">
<do type="accept">
<go href="#nextcard" />
</do>
<p>
Click the Accept Button</p>
</card>
```

```
<card id="nextcard" title="nextcard">
<p>
Thank you.
</p>
</card>

</wml>
```

In this example, the first card, `mycard`, prompts a single action: "Click the Accept Button."

The `<do>` tag indicates the type of action. The `"accept"` action turns one of the unit's soft keys into an Accept button. The user will see the word "Accept" over the button.

When the user clicks the Accept button, the second card appears. The command `<go href="#nextcard"/>` tells us that upon execution, a new card, called `nextcard`, will appear.

When `"nextcard"` appears, it will display the message `Thank you`. That is all it will do.

Summary

In this chapter, you learned about the wireless Web, what is involved in creating Web sites that work well in the wireless environment, and the various approaches to developing wireless-friendly Web sites. Don't forget to check out the comprehensive appendix (Appendix A), which includes Web links, and Appendix B, a glossary, and the tear-out card at the beginning of the book that pinpoints book chapters and segments according to your immediate area of interest.

Appendix A

Wireless Device and Technology Web Links

This appendix provides a list of links pertaining to the topics covered in this book, although the list is by no means comprehensive.

The vendors, regulatory bodies, and technology developers referred to in this book provide vast educational resources for students of wireless technology. You are encouraged to explore your areas of interest, and check Web sites frequently for updates, since the wireless world changes more rapidly than virtual ink can dry.

AIRWAVE TECHNOLOGIES

`www.gsmworld.com` All GSM-related issues, E-learning, current information, and article archives

`www.gsmcoverage.co.uk/` Maps of GSM Coverage

`www.cdg.org` Information about CDMA technology and implementation

`www.3gpp.org` Third Generation Partnership Project, a resource for all things 3G

`www.uwcc.org` Universal Wireless Communications, international cheerleaders for EDGE and TDMA-based 3G deployment, with many articles and E-learning resources

`www.mobilegprs.com` All matters pertaining to Mobile GPRS

`www.wow-com.com/internet/coverage/index.cfm?ID=62&SearchSection=&SearchCriteria=maps` Maps of wireless coverage access. If this link is too specific, try `www.wow.com.com`.[MBK1]

CELL PHONE PRODUCTS AND TECHNOLOGY

`www.ericsson.com` The Ericsson home site

`http://b41.sprintpcs.com` The Sprint PCS home site

`www.samsung.com` The Samsung home site

`www.javamobiles.com/index.html` Comprehensive information about Java-run phones and PDAs

`www.siemens.com` The Siemens home site

www.nextel.com/phone_services/directconnect.shtml The Nextel Direct Connect page

www.nokia.com/main.html The Nokia home site

EDUCATION — NETWORKING

www.informit.com Articles, information and E-learning on any topic even remotely related to the world of IT

www.protocols.com The Israeli network/telecommunications giant RAD sponsors this site, which comprehensively describes network protocols and provides vast E-learning opportunities.

EDUCATION — WIRELESS TECHNOLOGY

www.wirelessweek.com Articles and information about all things wireless

www.wsdmag.com The Wireless Systems Design magazine, which describes cellular technologies and related issues

www.networkmagazine.com A behemoth resource for wired, fixed wireless, and mobile wireless information

www.watmag.com The Wireless Access Technologies site, which provides comprehensive international news on wireless telecommunication issues, as well as E-learning resources

www.wow-com.com The World of Wireless Communications site, a wireless device comprehensive information source

http://searchnetworking.techtarget.com/ Comprehensive Educational Resources on Wireless Technology

E-MAIL NOTIFICATION AND PAGER PRODUCTS

www.rim.net The Research in Motion home site

ENTERPRISE APPLICATIONS

http://avantgo.com/frontdoor/index.html The AvantGo home site

http://appforge.com The AppForge home site

www.wapaka.com/wapaka/default.asp The Wapaka home site

www.thinairapps.com The ThinAirApps home site

www.handango.com The Handango home site

www.ameranth.com The Ameranth home site (Pocket PC 2002 applications partner)

FIXED WIRELESS TECHNOLOGY

`www.networkmagazine.com` A behemoth resource for wired, fixed wireless, and mobile wireless information

`www.wcai.com` A worldwide information source on fixed wireless

HOME RF

`www.homerf.org` Provides information and vast product info and links related to the Home RF

`http://commerce.motorola.com/consumer/QWhtml/home.html` The Motorola consumer wireless home site

`www.cayman.com/home_html.asp` The Cayman Systems home site (Home RF gateways)

LAN PRODUCTS AND TECHNOLOGY

`www.wlana.com` An industry-sponsored comprehensive site for all WLAN-related issues and technology

`www.bluetooth.com` A site about all things Bluetooth

`www.proxim.com` The Proxim home page (wireless networking products)

`www.linksys.com` The Linksys home site (networking products, especially wireless)

M-BUSINESS NEWS AND TECHNOLOGY

`www.mbizcentral.com` The M-Business Daily Web page, with information on all issues pertaining to wireless business

`www.sybase.com/products/mobilewireless` Sybase's comprehensive collection of mobile database products

`www.sap.com/solutions/mobilebusiness/` German e-business giant SAP's mobile business solutions site

MOBILE INTERNET

`www.wapforum.org` The official site of the WAP Forum

`www.gelon.net` Create WAP sites, simulate the appearance of a WAP site on your desktop computer

`www.palmos.com/dev/tech/webclipping/` Palm's official Web Clipping Application Development site

`www.w3.org/TR/1998/NOTE-compactHTML-19980209/` A W3C abstract that describes C-HTML in detail

PDA PRODUCTS AND TECHNOLOGY

www.pdabuzz.com News, resources, and hot-topic discussions on PDA-related issues

www.palm.com/products/palmvii/wireless.html The Palm.net Wireless Communication site for proprietary information on Palm wireless access options

www.palm.com The Palm home site

www.handspring.com The Handspring home site

www.sonystyle.com/micros/clie/ The Sony Clie home site

www.palmblvd.com/index.html PDA software, products, news, and views

www.palmgear.com PDA gear

www.nettechinc.com/palm.htm PDA resources for lawyers

http://freewarepalm.net Software for the Palm OS

www.zdnet.com/downloads/pilotsoftware Software for the Palm OS

www.palmpilotware.com Software for the Palm OS

www.purepalm.com/site/ PDA software, products, news, and views

www.pdamd.com/vertical/home.xml Palm resources for doctors

www.palmpower.com *Palm Power* magazine

www.edteck.com/palm/ Palm resources for professionals, especially educators

POCKET PCS AND HANDHELDS

www.microsoft.com/mobile/pocketpc/default.asp The Microsoft Pocket PC site

www.microsoft.com/mobile/pocketpc/pocketpc2002/default.asp The Microsoft Pocket PC OS 2002 site

http://products.hp-at-home.com/sub_category/sub_category.php?id=9 The Hewlett Packard Jornada site

www.compaq.com/products/handhelds/pocketpc/ The Compaq iPaq site

www.casio.com/personalpcs/section.cfm?section=19 The Casio Pocket PC home site

www.csd.toshiba.com/pda/pda_home.html The Toshiba handhelds home site

www.colorgraphic.net/cf/navigvoyagercf.html The ColorGraphic CF VGA adapter site

www.abrandnewworld.com A Brand New World, home of the GISMO

REGULATIONS AND STANDARDS

www.fcc.gov/wtb/auctions Information about the FCC and spectrum licensing issues

www.itu.int/home/index.html The International Telecommunications Union home page, which includes recommendations on spectrum usage and standards

http://standards.ieee.org/ The Institute of Electrical and Electronics Engineers (IEEE) standards site, for broad information about the 802.11b and other networking standards

SERVICE PROVIDERS

www.omnisky.com The OmniSky Web site

www.goamerica.com The Go-America home site

SUNDRY PRODUCTS

www.travroute.com The TravRoute site (GPS navigation products)

www.colorgraphic.net/cf/navigvoyagercf.html The ColorGraphic CF VGA adapter site

www.pentaxtech.com/Products/PJ200/PJ200_frameset.html The Pentax portable printer site

WAN PRODUCTS AND TECHNOLOGY (WIRELESS MODEMS)

www.socketcom.com/product/dpc.htm The Socket Digital Phone Card site

www.ositech.com The Ositech home site

www.novatelwireless.com/home_html.html The Novatel home site (Minstrel modems)

www.sierrawireless.com The Sierra Wireless home site

www.orinocowireless.com The Orinoco home site (wireless connectivity for mobile computers)

www.enfora.com The Enfora site (wireless connection devices)

Appendix B

Glossary

1G An abbreviation for First-Generation Wireless. 1G systems used analog encoding and circuit switching. They provided low-quality audio and slow, unreliable, and insecure data transfers. 1G systems are gradually being replaced by 2G and newer systems. AMPS and NMT are examples of 1G standards. Many mobile phone services in the United States are still 1G.

2G An abbreviation for Second-Generation Wireless. 2G systems use digital encoding and carry both voice and data. They provide some additional services, such as SMS. 2G systems are widely used today. GSM and TDMA are examples of 2G standards.

2.5G An example of economic reality colliding with technologic possibility. 2.5G is a way to make some of the features of 3G systems available to users, without requiring the industry to make massive investments as it moves from 2G to 3G networks. GPRS and EDGE are examples of 2.5G standards.

3G An abbreviation for Third-Generation Wireless. 3G systems use digital encoding and provide data rates in the millions of bits per second (Mbps). Some 3G systems use packet switching to more efficiently route data. With data rates in this range, 3G systems can support full-motion video, high-speed Internet access, and video conferencing. At the time of this writing, 3G systems are just beginning to appear outside the laboratory. WCDMA and UWC-136 are examples of 3G standards.

3GPP An acronym for Third-Generation Partnership Project. An organization that exists to create globally applicable specifications for a 3G Mobile System (UMTS) based on evolved GSM and related technologies such as GPRS and EDGE.

3GPP2 An acronym for Third-Generation Partnership Project 2. An organization that exists to create globally applicable specifications for the CDMA 2000 3G system. It is also an evolved version of CDMA technologies, in particular, cdmaOne.

4G An abbreviation for Fourth-Generation Wireless. No exact definition of 4G exists, but when most people speak of 4G, they are referring to an ultra-high-speed, packet-switched, wireless network capable of transferring information at perhaps 100 Mbps — a rate far higher than even the best consumer broadband connections of today. OFDM is a proposed 4G standard.

8-PSK (Phase Shift Keying) Modulation A modulation technique that uses eight phase changes to encode data on the carrier wave.

802.11b An IEEE standard that supports wireless data rates of up to 11 Mbps. Products that fully comply with 802.11b can interoperate with like devices from other manufacturers. Wi-Fi is a trade name for devices that comply with the 802.11b standard.

A Band One half of the total radio frequencies available for use within a given area. B Band is the other half. Each half is normally allotted to a different company providing cellular services in the area.

A-GPS An acronym for Assisted GPS. A hybrid mobile positioning system that supplements standard GPS signals with signals from Location Measuring Units (LMUs) that are integrated into the wireless network.

Access Point A wireless transceiver that connects wireless devices into a local area network (LAN).

Active-Matrix Display A display where each pixel includes its own driver transistor (or transistors) and associated electronics.

Ad Hoc Network A wireless network that does not contain an access point. Equipment from many WLAN vendors can function in an ad hoc network.

Advanced Mobile Phone System See *AMPS*.

Advice of Charge See *AOC*.

Air Interface The radio frequency portion of the wireless connection between two wireless devices, or between a wireless device and a base station. The air interface is usually the most troublesome portion of the circuit, as it is subject to interference and easily intercepted.

AMPS An acronym for Advanced Mobile Phone System. This 1G standard was the original standard used for mobile phones across much of the world, and is still in wide use in the United States. It operates in the 800–900 MHz frequency band.

Analog Encoding A method of encoding data that stores the data as a continuously varying quantity. AM and FM radio are standards that use analog encoding.

Anchor MSC A GSM term for the MSC that became responsible for a particular subscriber's information when that subscriber's mobile equipment established its most recent connection to the GSM network. The anchor MSC maintains responsibility for most of the subscriber's information and functionality, even if the mobile equipment subsequently moves into a cell controlled by a different MSC.

Anoto Pattern A pattern of dots printed on paper with slight offsets from a perfect grid pattern. The pattern is designed so that it could cover a piece of paper millions of square kilometers in area.

AOC Advice of Charge. This supplementary GSM service provides a subscriber with an estimate of the charges for a particular call.

API Application Program Interface. A collection of software functions packaged as binary libraries, allowing programmers to utilize development-enabling features.

ARIB An acronym for Association of Radio Industry Businesses. See *IMT-2000*.

Assisted GPS See *A-GPS*.

Association of Radio Industry Businesses See *IMT-2000*.

AUC An acronym for Authentication Center. A component of a GSM network that provides authentication and encryption information used to confirm a user's identity and ensure the confidentiality of a call.

Authentication Center See *AUC*.

B Band One half of the total radio frequencies available for use within a given area. A Band is the other half. Each half is normally allotted to a different company providing cellular services in the area.

Bandwidth A range of frequencies that contains a radio signal.

Base Station Controller See *BSC*.

Base Station Subsystem See *BSS*.

Base Transceiver Station See *BTS*.

BCCH An acronym for Broadcast Control Channel. One of the GSM control channels that is used to broadcast information such as the BTS identity frequency allocations (which mobile equipment is using which channel) and frequency-hopping sequences.

Bearer Services A term for GSM services that involve data, rather than voice.

Binary Runtime Environment for Wireless See *BREW*.

Blank-and-Burst A technique for sending commands along the AMPS voice channel. For a fraction of a second, the user's audio is muted (blanked) and a burst of data is sent.

Bluetooth A short-range, low-power wireless technology that works on the 2.5 GHz band and will serve a variety of usage profiles, as defined in the Bluetooth specification. It is also an open specification for reliable and secure short-range wireless communication (voice and data). Bluetooth replaces the cables that interconnect equipment in an office, allows a wireless device to synchronize with a desktop PC without any physical connection, and automatically forms and joins ad hoc personal networks. The initial specification sets a data transfer rate of up to 1 Mbps with a range of approximately 10 meters (about 30 feet). Bluetooth operates in the 2.4 GHz frequency band.

Bluetooth Profile A basic capability of a Bluetooth device—for example, service discovery, synchronization, or file transfer.

Bluetooth Usage Model Models of specific ways users could use their Bluetooth devices. Examples of usage models include the automatic synchronizer and the interactive conference.

BOQAM An acronym for Binary Offset Quadrature Amplitude Modulation. A variant of QAM.

Broadband PCS A particular class of digital wireless services that works in the 1900 MHz band of the electromagnetic spectrum. When people refer to PCS, they are normally referring to broadband PCS.

Broadcast Control Channel See *BCCH.*

Bps Bits per second. A measure of the amount of digital data a communication channel can carry. Today, long-range wireless systems (mobile phones, Palm VIIs, and so on) typically transfer thousands of bits per second (Kbps). Short-range wireless systems (Bluetooth and 802.11b networks) typically transfer millions of bits per second (Mbps).

BREW An acronym for Binary Runtime Environment for Wireless. A software platform that runs on mobile devices that use CDMA technology from Qualcomm. It allows applications written for BREW to run on a range of devices without needing to be customized for each device.

BSC An acronym for Base Station Controller. In a GSM network, each BSC controls one or more Base Transceiver Stations and connects them to the Mobile Services Switching Center responsible for the local service area.

BSS An acronym for Base Station Subsystem. As one of the components of a complete GSM system, the BSS handles communication between mobile stations (wireless handsets and other mobile devices) and the rest of the GSM system.

BTS An acronym for Base Transceiver Station. The transmitting and receiving equipment located at a cell site that is used to communicate with wireless devices within cells.

Burst Period A GSM term for the basic time frame in which a transmitter can send a particular user's data. The duration of a burst period is 0.577 milliseconds.

C-HTML Also CHTML. An acronym for Compact HTML. A variant on HTML designed to allow the display of Web pages on the screen of a mobile phone. It is supported by i-mode phones and used by i-mode content providers. Also written *CHTML.*

CAM An acronym for Constant Awake Mode. When power consumption is not an issue (for desktop PCs), CAM is the preferred mode of operation for a wireless network card. In this mode, the wireless transceiver is always powered. Although more power is used than if the transceiver were turned off, devices can respond to messages much more quickly.

Carrier Sense Multiple Access with Collision Avoidance See *CSMA/CA*.

Carrier Wave A continuous wave with known characteristics that can be modulated with a signal and transmitted. When received, the combined signal is demodulated, allowing the signal to be recovered.

CCK An acronym for Complimentary Code Keying. CCK modulation is applied to the signal in an 802.11 network when it is moving data at 5.5 Mbps and 11 Mbps.

CDMA An acronym for Code Division Multiple Access. This 2G technology uses digital encoding and spread spectrum transmission techniques to provide high-quality signals.

CDMA 2000 An acronym for Code Division Multiple Access 2000. This 3G technology evolves CDMA to higher speeds, moving it incrementally to 2 Mbps.

cdmaOne A name for the original 2G CDMA specification published by the ITU. The Sprint PCS service is built on cdmaOne. It supports a data rate of 14.4 Kbps, operating on either the 800 MHz band or the 1.9 GHz band.

CDPD An acronym for Cellular Digital Packet Data. An open wireless transmission technology that breaks digital data into packets and transmits it using idle network capacity caused by pauses in speech, gaps between conversations, and so on. CDPD works in the 800–900 MHz frequency band and allows for two-way 19.2 Kbps data transfer on AMPS cellular systems.

Cell A geographic area covered by multiple cell sites. Each cell is comprised of the region covered by (typically) three cell sites, situated along the edges of the cell.

Cell Broadcast A GSM capability that allows user to send a short (93-character) text message to everyone within a particular GSM cell.

Cell Cluster A group of cells within which no channels are reused.

Cell Global Identity and Timing Advance See *CGI+TA*.

Cell Site The point from which signals are broadcast into a cell. Each cell site is located on the edge of cells (typically three), not in the center.

Cell Splitting The process of dividing one existing cell into several smaller cells, allowing the cellular system to support more users in a given geographic area.

Cellular Radio The basic technology that underlies mobile phone systems. Cellular radio divides the coverage area into cells, each of which is covered by a single base station. Each cell uses a different frequency than its neighbors, with frequencies reused in cells that are far enough apart to avoid interference. The size of a cell is determined by the geography of the area covered by the cell, the power of the transceiver, and the predicted number of users.

Cellular Telecommunications and Internet Association See *CTIA*.

Cellular Telephone A term for a wireless telephone that communicates through a cellular radio network.

CGI An acronym for Cell Global Identity. In GSM systems, the CGI is a number assigned to every cell for the purpose of identifying that cell.

CGI+TA An acronym for Cell Global Identity and Timing Advance. The CGI+TA is a GSM network-based mobile positioning system that estimates the position of a mobile terminal within a given cell based on the time delay in a signal between the terminal and the base station.

Channel A range of frequencies used to carry a single connection.

Channel Pair Two channels in a wireless system, one for signals from a wireless device, the other for signals to the wireless device.

Channel Pollution A problem that occurs with CDMA-based systems when, for example, a handset can communicate with several base stations but none of the stations is clearly stronger.

Chip Rate The bit rate of the pseudorandom code used in a CDMA system.

CHTML Also C-HTML. An acronym for Compact HTML. CHTML is a subset of HTML that has been extended by NTT DoCoMo to apply to cellphones. The wildly popular DoCoMo i-mode service uses CHTML.

Ciphertext The encrypted payload (data plus CRC) in an encryption scheme.

Circuit-Switched Cellular Digital Packet Data See *CS CDPD*.

Circuit-Switching A technique for routing messages that entails setting up and maintaining a dedicated circuit between the end points. This circuit remains open for the duration of the communication. Circuit-switched systems have limited capacity, and users typically get charged by the length of time they are connected.

CLNP An acronym for ConnectionLess Network Protocol. CLNP provides basically the same features as IP.

Co-Channel Interference Interference that occurs when two or more signals are simultaneously transmitted on the same channel.

Code Division Multiple Access See *CDMA.*

Code Division Multiple Access 2000 See *CDMA 2000.*

Compact HTML See *CHTML.*

Complimentary Code Keying See *CCK.*

Constant Awake Mode See *CAM.*

Content Provider A company that provides the services users connect to with their wireless devices. News, traffic, weather, online shopping, or games and Internet access are some of the services delivered by content providers.

Control Channel A channel in a cellular system that is used to manage the activities of the system, instead of carry user signals.

Coverage Area The geographic area within which a wireless system can reliably function. Outside the coverage area, wireless devices cannot directly connect to their home network, although they may be able to connect indirectly by way of roaming.

CRC An acronym for Cyclic Redundancy Check. CRC is a form of error correction code.

CS CDPD An acronym for Circuit-Switched Cellular Digital Packet Data. CS CDPD is a variant on CDPD that allows devices to establish a dedicated CDPD connection for channels where the data traffic is expected to be very heavy.

CSMA/CA An acronym for Carrier Sense Multiple Access with Collision Avoidance. CSMA/CA is a technique designed to avoid data collisions on a wireless network. In a data collision, two stations transmit simultaneously, preventing either message from getting through. CSMA/CA is the collision-avoidance technique specified in the 802.11 wireless specification.

CTIA An acronym for Cellular Telecommunications and Internet Association. CTIA represents the interests of the wireless communications community before government regulators and policy makers.

Cyclic Redundancy Check See *CRC.*

D-AMPS An acronym for Digital-Advanced Mobile Phone Service. This 2G standard is a hybrid: it has a digital component based on TDMA, yet is also compatible with the large installed base of mobile phones using AMPS. D-AMPS operates in the 800 MHz and 1.9 GHz frequency bands. Also written DAMPS.

dBi An acronym for Decibels-Isotropic. In the expression of antenna gain, it is the number of decibels of gain of an antenna referenced to the zero dB gain of a free-space isotropic radiator — a "perfect" antenna.

DCS 1800 An acronym for Digital Communications System 1800. Another name for GSM operating in the 1800 MHz (1.8 GHz) frequency band. Also known as GSM 1800 or PCN. See *GSM.*

Decryption The conversion of data from an unreadable (encrypted) format into one that can easily be read.

Demodulation Any of the possible techniques for extracting a user signal from a carrier signal.

Denial of Service Attack See *DOS Attack.*

DHCP An acronym for Dynamic Host Configuration Protocol. DHCP automatically issues IP addresses to devices on a network. The devices retain these IP addresses for a specified length of time, then must release them back into the general pool maintained by DHCP. Most operating systems now provide DHCP.

Digital Encoding A method of encoding data that stores the data as a series of binary digits. GPRS and 802.11b are standards that use digital encoding.

Direct Sequence Spread Spectrum See *DSSS.*

Discontinuous Reception A method that calls for a receiver to turn itself off whenever it is not time to receive a burst of data.

Discontinuous Transmission A method that calls for a transmitter to turn itself off whenever there is silence in the conversation. This conserves significant power, as each person typically talks only 40 percent of the time in a typical voice call.

DOS Attack An acronym for Denial of Service attack. A type of attack that is designed to disable a computer network or a Web site by flooding them with Internet traffic.

DSSS An acronym for Direct Sequence Spread Spectrum. One of two types of spread spectrum radio transmission in use. DSSS spreads the signal across a wide frequency band.

DTMF An acronym for Dual Tone Multi-Frequency. A system of tones used in touch-tone telephone dialing.

Dual Band Device A wireless device that can connect to a particular network using two different frequency bands. For example, GSM operates on the 900 MHz band in some areas, and the 1800 MHz band (GSM 1800) in others. A dual band GSM device could connect in either area.

Dual Mode Device A wireless device that can connect to two different networks. Dual mode is typically used in the United States, where large areas have only analog AMPS (1G) phone coverage. A dual mode phone could use its native 2G service where available, and drop to 1G mode where 2G was not available.

Dual Tone Multi-Frequency See *DTMF.*

Dynamic Host Configuration Protocol See *DHCP*.

E-OTD An acronym for Enhanced Observed Time Difference. A hybrid mobile positioning system that works by measuring the observed time difference for signals from nearby base transceiver stations.

E-TACS An acronym for Extended Total Access Communications System. An extension of TACS, this 1G system is in use in parts of Europe and Asia.

EDGE An acronym for Enhanced Data rates for GSM and TDMA/136 Evolution (as currently defined by the ITU). This 2.5G standard increases the network capacity and data rates of existing GSM networks. It is expected to provide rates of up to 384 Kbps, and will allow network operators to provide some services comparable to those promised by 3G systems, without the expense of building an entirely new network.

EIR An acronym for Equipment Identity Register. A database of every piece of mobile equipment registered on a GSM network. Mobile equipment marked invalid in the EIR is prevented from using the network.

Electron Transparent Layer See *ETL*.

Electronic Ink One of the two or more substitutes for traditional ink that can be manipulated electronically. Electronic ink can be changed from light to dark by applying an electric field to a surface treated with the ink. Electronic ink is expected to lead to highly readable displays that consume very little power and are flexible. Color versions are also in development.

Electronic Serial Number See *ESN*.

Encryption The conversion of data into an unreadable format. Encryption provides security for your data or voice messages by making it difficult or virtually impossible to read the message without the proper decryption information.

Energy Density The amount of energy a battery can store in a given volume.

Enhanced Data Rates for GSM and TDMA/136 Evolution See *EDGE*.

Enhanced Data Rates for GSM Evolution See *EDGE*.

Enhanced Observed Time Difference See *E-OTD*.

EPOC A mobile device operating system championed by Symbian. EPOC competes with the Palm OS and Pocket PC, and is most popular in Europe.

Equipment Identity Register See *EIR*.

ESN An acronym for Electronic Serial Number. A number that uniquely identifies a particular wireless device. Used with AMPS and other systems.

Ethernet The most common wired LAN technology defined in the IEEE 802.3 specification. Some organizations are working to create a wireless extension of Ethernet.

ETL An acronym for Electron Transparent Layer. A layer of organic material between the cathode and the Emitter Layer of an OLED.

ETSI An acronym for European Telecommunications Standards Institute. A non-profit organization that produces voluntary telecommunications standards for Europe and other regions of the world.

European Telecommunications Standards Institute See *ETSI*.

Extended Total Access Communications System See *E-TACS*.

Extensible Markup Language See *XML*.

External Handover A GSM term for a handoff that must be managed with the involvement of one or more MSCs.

FAA An acronym for Federal Aviation Administration. A United States government agency that is responsible for aviation issues. It is also developing the WAAS system for increasing the accuracy of civilian GPS signals.

Facsimile Group III An international standard for sending facsimile data to standard fax machines.

FCC An acronym for Federal Communications Commission. The U.S. government agency responsible for regulating telecommunications.

FDD An acronym for Frequency-Division Duplexing. A technique for separating the signals going from handset to base station by transmitting them on different channels.

Federal Aviation Administration See *FAA*.

Federal Communications Commission See *FCC*.

FHSS An acronym for Frequency Hopping Spread Spectrum. One of two types of spread spectrum radio transmission in use. In FHSS, the transmitter and receiver synchronously skip from one narrow frequency band to another within the wide overall frequency band or the signal.

First-Generation Wireless See *1G*.

First Paging Channel The channel a wireless device initially monitors when looking for commands on the wireless system's control channels.

Flexible Organic Light Emitting Diode See *FOLED*.

FM An acronym for Frequency Modulation. A technology for encoding an analog signal on a carrier wave by varying the frequency of the carrier wave in proportion to the level of the signal. More technically, a method of impressing data onto an alternating-current (AC) wave by varying the instantaneous frequency of the wave. This scheme can be used with analog or digital data.

FOLED An acronym for Flexible Organic Light Emitting Diode. A type of organic light-emitting diode that could be applied to curved surfaces, bent or even folded up for storage. See *OLED*.

FOMA An acronym for Freedom Of Mobile Multimedia Access. A brand name for NTT DoCoMo's 3G mobile communication service. The service is based on WCDMA, and trials began in May of 2001.

Forward Channel The communication channel in wireless systems that carries data from the base station to the mobile terminal.

FPLMTS An acronym for Future Public Land-Mobile Telephone Systems. A 3G spec from the ITU. FPLMTS is now known as IMT-2000. See *IMT-2000*.

Freedom Of Mobile Multimedia Access See *FOMA*.

Frequency-Division Duplexing See *FDD*.

Frequency Hopping The act of regularly changing the frequency of a transmission, typically to minimize interference.

Frequency Hopping Spread Spectrum See *FHSS*.

Frequency Modulation See *FM*.

Future Public Land-Mobile Telephone Systems See *IMT-2000*.

Gateway GPRS Support Node See *GGSN*.

Gaussian Minimum Shift Keying See *GMSK*.

General Packet Radio Service See *GPRS*.

GGSN An acronym for Gateway GPRS Support Node. A component of a GPRS-equipped network that provides the interface between the network and external IP or X.25 networks.

GHz An acronym for gigahertz. A frequency of billions of cycles per second.

GIWU An acronym for GSM Interworking Unit. The GIWU gives a GSM system an interface to various data networks.

Global Positioning System See *GPS*.

Global System for Mobile Communications See *GSM*.

GMSC An acronym for Gateway Mobile Services Switching Center. When a GSM system's Mobile Services Switching Center contains a gateway to another network, it is called a GMSC.

GMSK An acronym for Gaussian Minimum Shift Keying. A modulation/demodulation technique that applies Gaussian filtering to a continuous-phase frequency-modulated carrier wave.

GPRS An acronym for General Packet Radio Services. A 2.5G technology that enhances existing GSM systems, adding packet-switching capabilities, including a continuous connection to the network, the ability to charge by the bit, instead of by connect time. With GPRS, a GSM system could offer data rates of 20 to 30 Kbps.

GPS An acronym for Global Positioning System. A terminal-based mobile positioning system that computes the position of the receiver based on a measure of the length of time it takes for signals to travel to the device from a constellation of GPS satellites in low earth orbit.

Group-Graphics Some mobile phones allow users to group callers, and display a graphic on their mobile phone when someone from a particular group calls.

GSM An acronym for Global System for Mobile Communications. This 2G standard is currently the world's most widely used digital mobile wireless system (used by over half a billion people in over 169 countries), and the de facto standard across Europe. It provides digital voice, data, and text messaging (SMS) services. GSM uses TDMA technology and operates on 800 MHz, 1.8 GHz, and 1.9 GHz frequency bands. Unenhanced GSM delivers a data rate of 9.6 Kbps.

GSM 1800 An acronym for Global System for Mobile Communications, 1800 MHz (1.8 GHz). See *GSM*.

GSM Interworking Unit See *GIWU*.

Handoff The transition that occurs when a cellular user moves from one cell to another, and a new station takes responsibility for the transmission.

Handover Another term for *handoff*, commonly used outside of North America.

Hard Handoff A handoff that requires the handset to switch frequencies. Hard handoffs may cause audible effects as the handset switches from one frequency to another. They can also result in calls being dropped during the transition.

HCS An acronym for Hierarchical Cell Structure. A system whereby macro, micro, and pico (large, medium, and small) cells can all coexist. Micro and pico cells can exist within macro cells, with the mobile phone choosing to communicate with the most appropriate cell type for the current circumstances. This is similar to, but not the same as, cell splitting.

HDML An acronym for Handheld Device Markup Language. A version of HTML designed specifically for use on mobile phones, PDAs, and other devices with small display screens. HDML was developed by Unwired Planet (now a part of OpenWave Systems, Inc.).

Hidden Node A node of a wireless network that, due to its location or other conditions, cannot detect when another node is also communicating with the desired recipient. The result is frequent data collisions, lost messages, and greatly reduced performance.

Hierarchical Cell Structure See *HCS*.

HIL An acronym for Hole Injection Layer. A layer of organic material that contributes holes (positive charges) to an OLED.

HLR An acronym for Home Location Register. The most important database in a GSM system, which contains information for every user subscribed to the system.

Hole Injection Layer See *HIL*.

Hole Transparent Layer See *HTL*.

Home Location Register See *HLR*.

Home System Identification Number See *SID*.

HSCSD An acronym for High Speed Circuit Switched Data. A technology (expected to eventually become part of the GSM specification) that increases the data rate of GSM by using up to four sequential TDMA time slots for a single user. With HSCSD, each time slot can move data at up to 14.4 Kbps, with a maximum rate of 57.6 Kbps (when four sequential time slots are used).

HTL An acronym for Hole Transparent Layer. A layer of organic material between the HIL and the Emitter Layer of an OLED.

HTML An acronym for HyperText Markup Language. The language used to create most Web pages. Some wireless devices can work with HTML, or a subset of it. These devices can display and interact with a much greater portion of the information on the Internet than devices that only work with HDML or WAP.

HTTP An acronym for HyperText Transfer Protocol. The underlying format used for transferring data on the World Wide Web. Wireless device manufacturers and standards bodies are working to create wireless HTTP standards.

i-Mode A proprietary 2G digital, packet-switched wireless voice and data system from NTT DoCoMo. Thanks to its packet-switched nature, an i-Mode phone can remain continuously connected to the network. It offers services from many content providers, and can send and receive e-mail within the system or across the Internet. This 9.6 Kbps system is wildly popular in Japan, and is beginning to spread to other parts of the world.

iCalendar A set of data formats and protocols used for online calendars and schedules, published by the IETF.

iDEN An acronym for Integrated Digital Enhanced Networks. A proprietary technology from Motorola that combines a digital cellular phone, two-way radio, alphanumeric pager, and data/fax modem into a single device. iDEN systems are in operation in more than a dozen countries. iDEN is based on GSM.

IEC An acronym for International Engineering Consortium. An organization that, among its many other activities, provides some excellent Web-based tutorials at their Web Pro Forum site.

IEEE An acronym for the Institute of Electrical and Electronics Engineers. Among the many functions of this organization is the establishment of standards such as the 802.11b wireless networking standard.

IETF An acronym for Internet Engineering Task Force. An open community of people concerned about the evolution of Internet architecture, as well as the smooth operation of the Internet.

IM An acronym for Instant Messaging. A technology that allows two or more people to have a real-time conversation by sending text messages to each other. IM systems typically know when users are connected to the system, and can notify users that their friends are online. There are several different, and incompatible, IM standards in widespread use today. Efforts are also underway to make it possible for IM systems to exchange messages with each other.

IMEI An acronym for International Mobile Equipment Identity. A unique IMEI is assigned to each unit of GSM mobile equipment when it is manufactured. This number is stored in a database and can be marked invalid if the unit is stolen or approval for this type of device is revoked.

IMT-2000 An acronym for International Mobile Telecommunications-2000. IMT-2000 is a 3G standard developed by the ITU. Devices supporting IMT-2000 can communicate with a variety of telecommunication networks, providing worldwide coverage. IMT-2000 was previously known as FPLMTS. In Europe, IMT-2000 is called UMTS. In Japan, it is called J-FPLMTS.

Infrared See *IR*.

Infrared Data Association See *IrDA*.

Initialization Vector See *IV*.

Instant Messaging See *IM*.

Institute of Electrical and Electronics Engineers See *IEEE*.

Integrated Digital Enhanced Networks See *iDEN*.

Internal Handover A GSM term for a handoff that can be managed by a single BSC, without the involvement of an MSC.

International Engineering Consortium See *IEC.*

Internet Engineering Task Force See *IETF.*

Internet Protocol See *IP.*

IP An acronym for Internet Protocol. This protocol defines the format of data packets that travel on the Internet; it also specifies the addressing scheme used.

IP Multicast The ability to broadcast data from one source to many destinations, using IP.

IPng An acronym for Internet Protocol Next Generation. See *IPv6.*

IPsec An acronym for IP Security. The security protocol used by IPv6.

IPv4 An acronym for Internet Protocol Version 4. The version of IP that currently serves the Internet.

IPv6 An acronym for Internet Protocol Version 6. A proposed replacement to the current IP standard, Ipv4. Version 6 vastly expands the address space that can be reached using IP. Also known as *IPng*.

IR An acronym for Infrared. A portion of the electromagnetic spectrum used by some mobile devices for short-range communication.

IrDA An acronym for Infrared Data Association. A group of hardware manufacturers. Also a standard for transmitting data using IR. Some printers, PDAs, and PCs come with IrDA ports. Bluetooth is likely to eventually replace IrDA in most applications.

IrOBEX See *OBEX.*

IS-95a The first CDMA standard, known as cmdaOne. Provides data rates of up to 14.4 Kbps.

IS-95b An upgrade from IS-95a, increasing maximum data rates to 115.2 Kbps.

ISM An acronym for Industrial, Scientific, and Medical, representing a portion of the spectrum that is set aside for unlicensed use. The 2.4 GHz ISM band is used by Bluetooth, IEEE 802.11, and other wireless technologies.

ITU An acronym for International Telecommunications Union. An international organization through which governments and the private sector coordinate global telecom networks and services. The ITU is a United Nations agency.

IV An acronym for Initialization Vector. A string of bits used by a string cipher to begin encrypting data. The RC4 cipher used by the WEP standard has a 24-bit initialization vector.

Japanese Total Access Communications System See *JTACS*.

Java An object-oriented language created by Sun Microsystems and formally released to the world in 1996. It can be used to create small applications (applets) that will run on many different operating systems, making it useful when developing software for diverse types of hardware. Wireless devices that support Java are beginning to appear.

Java Bytecode The platform-independent code that is created by a Java compiler. It should run on any hardware that has an appropriate *Java Virtual Machine*.

Java Virtual Machine An interpreter or runtime environment that supports Java. Having the appropriate Java virtual machine installed allows a wireless device to run Java.

JFPLMTS See *IMT-2000*.

Joint Detection A matrix calculation technique that greatly improves the efficiency of CDMA signal detection by eliminating the effect of their mutual interference. Applicable only to situations where the total number of signals is small.

JTACS An acronym for Japanese Total Access Communications System. A variant of TACS that was used in Japan.

Kbps Thousands of bits per second. See *Bps*.

Keyschedule The combination of a WEP key and an initialization vector that is used to encrypt data on 802.11b LANs.

Keystream A string of bits generated from a keyschedule, equivalent in length to the data that will be transmitted, plus a CRC.

Kph Kilometers per hour.

LA An acronym for Location Area. In a GSM system, the LA is a group of cells controlled by a single MSC. When the system pages a user, it sends the page to the entire LA.

LAI An acronym for Location Area Identity. The LAI number specifies a particular Location Area within a GSM network.

Last Paging Channel The highest channel number that a wireless device monitors when looking for commands on the wireless system's control channels.

LCD An acronym for Liquid Crystal Display. A reflective display technology used in most mobile devices today. Most commercially available LCDs are hard to read in direct sunlight and require a backlight to be readable at night.

Line of Sight An unobstructed straight path between two transceivers. A clear line of sight is usually required for long-range directional transmissions, which are found in some wireless networks.

Liquid Crystal Display See *LCD*.

LMU An acronym for Location Measurement Unit. A facility added to wireless networks that provides supplemental GPS data that assists GPS devices in determining accurate positions when GPS satellite coverage is limited. Also used for a facility added to networks to compute the uplink time of arrival in networks that use the UL-TOA positioning system.

Location Area See *LA*.

Location Measurement Unit See *LMU*.

MAC An acronym for Media Access Control. A unique address assigned to every network interface device by the manufacturer.

Macro Cell The largest type of cell in the HCS system supported by D-AMPS.

Malware Any kind of software that is harmful to the hardware or software of the device it runs on. Viruses and worms are examples of malware.

Markup Language A coding system used to structure, index, and link text files. HTML and WML are two examples of markup languages that wireless devices work with.

Mbps Millions of bits per second. See *Bps*.

ME An acronym for Mobile Equipment. The hardware necessary to connect to a GSM network. To ensure the functioning of the mobile equipment, a SIM card is also required.

Media Access Control See *MAC*.

MExE An acronym for Mobile Execution Environment. A standard that defines a framework for 2.5G and 3G mobile devices. It incorporates both WAP and Java. MExE currently defines three classmarks—technical requirements that a mobile device must meet to support certain technologies. A handset can support any number of MExE classmarks.

MHz Megahertz. A frequency of millions of cycles per second.

Micro Cell The intermediate-sized cell in the HCS system supported by D-AMPS. Micro cells can exist within macro cells.

MIN An acronym for Mobile Identification Number. A 34-bit representation of a wireless device's 10-digit telephone number.

Minimum Shift Keying See *MSK*.

Ministry of Posts and Telecommunications See *MPT*.

MIPv6 An acronym for Mobile Internet Protocol Version 6. A protocol added to IPv6 that allows mobile devices to remain authenticated with a network, even as they physically move through the network.

Mobile Equipment See *ME*.

Mobile Identification Number See *MIN*.

Mobile IPv6 See *MIPv6*.

Mobile Position Center See *MPC*.

Mobile Services Node See *MSN*.

Mobile Services Switching Center See *MSC*.

Mobile Station In a GSM context, the combination of mobile equipment (a mobile terminal, typically a phone) and a SIM card. When both elements are present, a mobile station is capable of connecting to an appropriate GSM network.

Mobile Subscriber Unit See *MSU*.

Mobile Telephone Switching Office See *MTSO*.

Mobitex® An all-digital, packet-switched technology designed for data-only transfer. Mobitex is based on GMSK technology developed by Ericsson. Over 30 Mobitex networks are currently in worldwide operation.

Modulation Any of the possible techniques for combining a signal with a carrier wave.

MPC An acronym for Mobile Position Center. In an UL-TOC mobile positioning system, the MPC is the facility that computes the location of a wireless terminal based on the time of arrival of a signal from the mobile terminal.

MPT An acronym for Ministry of Posts and Telecommunications. The Japanese government agency that regulates telecommunications.

MSC An acronym for Mobile Services Switching Center. The MSC communicates with one of more Base Station Controllers, as well as telephone and data systems that are not part of the GSM system. The MSC also provides the functionality needed to support mobile subscribers.

MSK An acronym for Minimum Shift Keying. A modulation technique that encodes data by varying the frequency of the carrier wave so that the difference between the frequency of a logical one and a logical zero is exactly one half the data rate.

MSN An acronym for Mobile Services Node. A component of a GSM system that provides the capability to deliver mobile intelligent network functionality such as mobile VPN service.

MSU An acronym for Mobile Subscriber Unit. An older term for mobile telephones and other devices that connect to cellular phone systems.

MTSO An acronym for Mobile Telephone Switching Office. A switch that connects a group of cells to the main phone system, authenticates callers, and handles the switching of users between cells as the users cross cell boundaries.

Multipath Fading A problem that occurs when portions of a radio signal reflect off objects near the transmitter and receiver, thus arriving at the receiver at slightly different times. Because these signals are slightly out of phase with each other, they reduce the usable signal strength at the receiver.

Narrowband PCS A portion of the radio spectrum near 900 MHz that is dedicated to PCS. Three MHz has been allocated to narrowband PCS in the 901–902, 930–931, and 940–941 MHz bands.

Near-Far Problem A spread spectrum situation in which the power of a signal received from a nearby spread spectrum transmitter overwhelms the power of a signal from a distant transmitter, resulting in the need for a huge spreading bandwidth so that the distant station can receive service.

Network and Switching Subsystem See *NSS*.

Network Operator A company with a license to provide wireless telephony services.

NMT An acronym for Nordic Mobile Telephone. One of the first 1G standards, developed and used in Denmark, Finland, Iceland, Norway and Sweden.

NMT 450 A version of NMT operating at 450 MHz. See *NMT*.

NMT 900 A version of NMT operating at 900 MHz. See *NMT*.

Nordic Mobile Telephone See *NMT*.

NSS An acronym for Network and Switching Subsystem. One of the components of a complete GSM system. The NSS handles communication between the GSM system and other communications systems (the wired telephone system, for example). The NSS also provides all the functionality needed to manage mobile subscribers. The primary component of the NSS is the Mobile Services Switching Center.

OASIS An acronym for Organization for the Advancement of Structured Information Systems. An industry organization dedicated to minimizing overlap in XML languages and XML language standard initiatives by providing public information to XML information and schema.

OBEX An acronym for Object Exchange. A set of high-level protocols that define the way data objects such as vCard contacts can be exchanged using IRDA (IrOBEX), Bluetooth, or other technologies. OBEX support is built into a variety of devices and operating systems, including Microsoft Windows 2000 and Palm OS–compatible devices.

OELD An acronym for Organic Electroluminescent Diode. See *OLED*.

OFDM An acronym for Orthogonal Frequency Division Multiplexing. A technology that has been proposed as part of a 4G standard. It offers several potential benefits for 4G systems, particularly the ability to deal effectively with multipath. There are several competing variations on OFDM and no unified standard.

Off-Hook An old telephone term that means the user has picked up the phone. Used in AMPS-based systems to tell the cellular network that the wireless device is ready to receive signals on the voice channel.

OLED An acronym for Organic Light Emitting Diode. A display technology that uses organic molecules instead of liquid crystals. The molecules emit light when stimulated, making them advantageous for use in displays. Work is being done on flexible displays (FOLEDs) and transparent displays (TOLEDs) using OLED technology. Sometimes also called OELD (Organic Electroluminescent Diode).

OMC An acronym for Operations and Maintenance Center. The facility from which a network operator monitors and controls a GSM network. The OMC is connected to all elements of the NSS and to the BSCs.

Omnidirectional Dipole Antenna An antenna that transmits and receives equally well in all directions.

On-Hook An old telephone term that means the user has not picked up the phone. Used in AMPS-based systems to tell the cellular network that the wireless device is not ready to receive signals on the voice channel.

Operation and Support Subsystem See *OSS*.

Operations and Maintenance Center See *OMC*.

Opt-In An approach to the privacy of location information that requires the user to actively choose to transmit location information, typically by pressing a button on their wireless device.

Organic Electroluminescent Diode See *OLED*.

Organic Light Emitting Diode See *OLED*.

Organization for the Advancement of Structured Information Systems See *OASIS*.

Orthogonal Frequency Division Multiplexing See *OFDM*.

OSS An acronym for Operation and Support System. One of the components of a complete GSM system. The OSS consists of an Operations and Maintenance Center and its connections to the Network and Switching Subsystem as well as to Base Station Controllers. The overall GSM system is operated and maintained through the OSS.

Packet The fundamental message unit in a packet-switched network. A packet typically contains user data, routing information, and sometimes error detection and correction data.

Packet Broadcast Control Channel See *PBCCH*.

Packet Common Control Channel See *PCCCH*.

Packet Data Channel See *PDCH*.

Packet Data Traffic Channel See *PDTCH*.

Packet-Switched A term that describes information (voice or data) that is broken up into small digital chunks (packets), each of which contains addressing and sequencing information, as well as a piece of the information that is to be transferred. Each packet is routed to its destination by whatever path is most appropriate at the instant the packet is sent. Devices that use packet switching do not tie up a dedicated communication circuit for the entire time they are connected, as do circuit-switched devices. From the user's perspective, packet-switched devices appear to be continuously connected. Users of packet-switched systems are typically charged by the packet, not by the minute.

Passive-Matrix Display A display where individual pixels do not have their own integrated driver electronics.

PBCCH An acronym for Packet Broadcast Control Channel. A channel that carries general system information being broadcast across the network.

PCCCH An acronym for Packet Common Control Channel. A channel that carries information on the imminent start of a packet transmission; it also notifies data terminals of the availability of additional time slots to transmit on.

PCN An acronym for Personal Communications Network. Another name for GSM operating in the 1.8 GHz frequency band. See *GSM*.

PCS An acronym for Personal Communications Services. A term, defined by the United States FCC, that refers to digital cellular technologies deployed in the U.S. PCS services are 100 percent digital and operate in the 1.9 GHz frequency band.

PDA Personal Digital Assistant.

PDCH An acronym for Packet Data Channel. A physical channel in a BTS that can carry GPRS packet data.

PDTCH An acronym for Packet Data Traffic Channel. A logical channel within the physical PDCH that carries the user data.

Personal Digital Assistant See *PDA*.

Personal Information Manager See *PIM*.

Phase Shift Keying See *PSK Modulation*.

Pico Cell The smallest-sized cell in the HCS system supported by D-AMPS. Pico cells can exist within macro and micro cells.

Piconet A communication channel through which two or more Bluetooth devices communicate with each other. Multiple piconets can coexist in a given area.

PIM An acronym for Personal Information Manager. A tool that holds a user's personal information, such as a calendar and a contact list.

PKI An acronym for Public Key Infrastructure. A global infrastructure that supports public key encryption. Such an infrastructure does not yet exist.

Power Save Mode See *PSM*.

Primary Cell A type of battery that is not rechargeable.

Provider Logos The graphics that identify the service provider for the network a mobile phone connects to. Provider logos appear on many brands of mobile phone whenever the device is turned on. Some phones allow users to replace the provider logo with one of their own choice.

Pseudo Random Code A binary code used in CDMA systems. When viewed in the frequency domain, a pseudo random code looks like noise. In CDMA systems, the pseudo random code is used to spread a digital signal across a much wider frequency band than it would otherwise occupy.

PSK Modulation An acronym for Phase Shift Keying Modulation. A family of techniques that encodes a digital signal on a carrier wave by changing the phase of that carrier wave.

PSM An acronym for Power Save Mode. In this mode, the transmitter in a wireless device is turned off to conserve power. Overall system performance is reduced when devices are in power save mode, since they must power up before transmitting.

PSTN An acronym for Public Switched Telephone Network. The overall worldwide telephone system, which includes the local networks, exchange area networks, and long-haul network.

Public Key Infrastructure See *PKI.*

Public Switched Telephone Network See *PSTN.*

QAM An acronym for Quadrature Amplitude Modulation. A modulation technique that encodes data using two amplitude-modulated sine waves of the same frequency, but 90 degrees out of phase, allowing allows a carrier wave to carry twice the data it would otherwise carry.

QOQAM An acronym for Quaternary Offset Quadrature Amplitude Modulation. A variation on QAM. See *QAM.*

QoS An acronym for Quality of Service. The ability to guarantee various levels of speed and reliability for data delivery across a network.

QPSK Modulation An acronym for Quadrature Phase Shift Keying Modulation. A form of Phase Shift Keying that uses four phase changes.

Quadrature Phase Shift Keying See *QPSK Modulation.*

Quality of Service See *QoS.*

Quaternary Offset Quadrature Amplitude Modulation A variant of QAM. See *QAM.*

Radio Transmission Technology See *RTT.*

Range A measure of the linear distance over which a wireless system can send and receive a signal.

RC4 The stream cipher used for WEP encryption in the 802.11b wireless standard.

Receiver Sensitivity A measure of the weakest signal that a receiver can detect and correctly extract data from.

Relay MSC A GSM term for the MSC that becomes responsible for a particular subscriber's information after that subscriber's mobile equipment comes under the control of an MSC different than the one it originally established a connection with. The relay MSC manages handovers for the subscriber's mobile equipment until such time as the ME moves into a cell controlled by a different MSC. The responsibility for most of the subscriber's information and functionality remains with the anchor MSC. Also called Serving MSC.

Reverse Channel The communication channel in some wireless systems that carries data from the mobile terminal to the base station.

RF An acronym for Radio Frequency. A generic term used to refer to radio-based technologies, as well as to electromagnetic waves with frequencies below 3 GHz.

RF Multipath An acronym for Radio Frequency Multipath. A condition caused when radio signals take slightly different paths to the receiver (due to reflection and interference by objects between the transmitter and receiver).

RF Spectrum An acronym for Radio Frequency Spectrum. A general term for the range of radio frequencies.

Ring Tones The tunes you sometimes hear when a mobile phone rings. Some mobile phones (primarily Nokia, but also certain Sagem and Motorola devices) allow the user to add custom ring tones, which are delivered via SMS right to the mobile phone.

Roaming The ability of some wireless networks to allow transceivers to physically move within the coverage area, without the transceiver losing a connection to the LAN. Also, the ability of some wireless telephone systems to allow the user to maintain coverage while out of the coverage area of the user's normal network provider.

RTT An acronym for Radio Transmission Technology. Another term for *Air Interface*.

SA An acronym for Selective Availability. An intentional error induced into the signals from GPS satellites to degrade their accuracy. SA has been disabled since May 2000.

SAT An acronym for Supervisory Audio Tone. An audio tone that is used in some cellular systems to distinguish between wireless devices that are on the same channel but in different cells.

Scatternet Two or more piconets coexisting at least partially in the same location at the same time.

SDR An acronym for Software Defined Radio. A radio transceiver that uses digital signal processors, high-speed microprocessors, and software to manage the functions of a data terminal that are traditionally handled by custom microelectronics.

Second Generation See *2G*.

Secondary Cell A type of battery that is rechargeable.

Selective Availability See *SA*.

Serving GPRS Support Node See *SGSN*.

Serving MSC See *Relay MSC*.

SGML An acronym for Standard Generalized Markup Language, the markup language from which such languages as HTML and XML are derived. SGML itself plays no real part in the story of wireless devices.

SGSN An acronym for Serving GPRS Support Node. A component of a GPRS-equipped network that provides authentication and mobility management for GPRS users.

SID An acronym for Home System Identification Number. A number that identifies the wireless service provider a wireless device belongs to. Also written SIDH.

SIDH See *SID*.

SIM An acronym for Subscriber Identity Module. A smart card inserted into all GSM handsets that identifies the user to the network, handles authentication, and stores basic user and network information.

Smart Antenna System A system that combines many antenna elements with powerful signal processing to create an "aim-able" antenna.

Smartphone The generic term for a wireless digital phone that has the ability to browse the Web, as well as receive and send short text messages and e-mail. Some Smartphones also provide features such as a calendar, scheduling, and other built-in PDA functions.

SMS An acronym for Short Message Service. A technology that allows users to send short (160 characters or less) text messages. These messages can be sent from a wireless handset or Web-based SMS Gateway to the network operator's message center. The message center delivers messages to the target handset as soon as possible, and stores them when the handset is not accessible.

Soft Handoff A term that describes the way CDMA systems hand off calls between base stations. The handset manages the handoff. Furthermore, it is not required to change frequencies, which greatly reduces the audible effects of handing off the call, as well as the chance that the call will be dropped.

Software Defined Radio See *SDR*.

Spatial Channels An application of smart antenna technology that allows wireless devices to be distinguished from one another based on their position. Just as frequency division multiple access systems create different frequency channels to distinguish between devices, a cellular system equipped with smart antennas could establish spatial channels, distinguishing between devices by their physical location.

Spread Spectrum A technology that spreads a signal across a much larger portion of the RF spectrum than it originally occupied. Spread spectrum provides more security and resistance to interference than other technologies.

SSID An acronym for Service Set ID. The network "name" of a wireless LAN access point in the 802.11b standard. Provides basic access control.

SSL An acronym for Secure Socket Layer. A network protocol designed for encrypted Internet transmissions.

Standard Generalized Markup Language See *SGML*.

Stream Cipher An encoding scheme that works on a stream of data.

Subscriber Identity Module See *SIM*.

Subscriber Services The set of services that GSM users can expect as part of their basic GSM subscription.

Supervisory Audio Tone See *SAT*.

Supplementary Services Services that GSM subscribers can add, for an additional fee, to their basic GSM subscription.

SyncML An acronym for Synchronization Markup Language. A standard for data synchronization that is optimized for wireless devices and networks. The goal of the organizations working on SyncML is to ensure that wireless networks can support synchronization of any data on the network. SyncML is based on XML. The standard is designed to function on all major transport layers, including HTTP, WSP, and OBEX.

TACS An acronym for Total Access Communication System. A 900-MHz, 1G standard that is based on AMPS. TACS was used in England, and eventually Japan and Hong Kong.

TD-SCDMA An acronym for Time Division Synchronous Code Division Multiple Access. A 3G technology (not currently approved by the ITU for IMT-2000) that adds aspects of TDMA to CDMA, along with various other innovations. Promoted by China and Siemens AG. Sometimes also called *TDD Low Chip Rate*.

TDD An acronym for Time-Division Duplexing. A technique for separating the signals going from handset to base station by transmitting them on the same channel, but at different times.

TDD Low Chip Rate See *TD-SCDMA*.

TDMA An acronym for Time Division Multiple Access. A 2G technology that allows multiple users to share a given frequency by allocating each a different time slot. GSM is one of the 2G standards that uses TDMA. TDMA is another name for *D-AMPS*.

TDOA An acronym for Time Difference of Arrival. The difference in the arrival time of a signal from a mobile terminal at a pair of Location Measurement Units. This value is used when calculating the position of a mobile data terminal by means of the UL-TOA technique.

Telecommunications Industry Association See *TIA*.

Telematics Wireless voice and data communications between an automobile and another location, usually a fixed information and service center.

Teleservices GSM services that are related to voice.

Third Generation Wireless See *3G*.

Third Generation Partnership Program See *3GPP*.

TIA An acronym for Telecommunications Industry Association. Created in 1988, this organization is responsible for developing telecommunication standards for use in the United States (other countries are free to adopt these standards as well). The TIA is responsible for the standards that govern such wireless technologies as AMPS and PCS.

Time Difference of Arrival See *TDOA*.

Time-Division Duplexing See *TDD*.

Time Division Multiple Access See *TDMA*.

Time Division Synchronous Code Division Multiple Access See *TD-SCDMA*.

TOLED An acronym for Transparent Organic Light Emitting Diode. A type of organic light-emitting diode in which both electrodes are transparent. Such a display could be applied to vehicle windscreens or head-mounted displays. See *TOLED*.

Total Access Communication System See *TACS*.

Transceiver A transmitter-receiver—in other words, a device that can both send and receive wireless signals.

Transparent Organic Light Emitting Diode See *TOLED*.

Transponder A transceiver that automatically responds to a signal it receives by sending out a signal of its own.

Transport Protocol An agreed-to set of rules for transferring data between two devices.

Tri-Band Phone A phone that can function on three distinct frequency bands (for example, a GSM phone that can work on 900 MHz, 1800 MHz, and 1900 MHz).

Triple DES An acronym for Triple Data Encryption Standard. A variant on the DES encryption algorithm that is three times slower but vastly more secure than its predecessors.

UL-TOA An acronym for Uplink Time of Arrival. A network-based mobile location system that computes the location of a mobile terminal based on the time it takes for a signal to travel from a mobile terminal to at least four measuring stations.

UMTS An acronym for Universal Mobile Telecommunications System. See *IMT-2000*.

UMTS Terrestrial Radio Access See *UTRA*.

Universal Mobile Telecommunication System An acronym for UMTS. See *IMT-2000*.

Universal Wireless Communication-136 See *UWC-136*.

Uplink Time of Arrival See *UL-TOA*.

UTRA An acronym for UMTS Terrestrial Radio Access. A term that is sometimes used to describe the ground-based (as opposed to satellite-based) radio transmission technologies.

UWC-136 An acronym for Universal Wireless Communication-136. This 3G standard provides an evolutionary path from AMPS to the latest technology. It employs TDMA, GPRS, and EDGE to gain speed and flexibility over earlier standards.

vCalendar A platform-independent standard for exchanging calendar and schedule information between PCs, PDAs, and other mobile devices.

vCard A platform-independent standard for exchanging calendar and schedule information between PCs, PDAs, and other mobile devices. vCard automates the transfer of the type of personal data you would find on a business card: e-mail address, phone number, and so on.

vMessage A platform-independent standard for exchanging e-mail and other messages between PCs, PDAs, and other mobile devices.

vNote A platform-independent standard for exchanging short notes between PCs, PDAs, and other mobile devices.

Vocoder Software that converts analog voice signals to digital form and back again.

Voice Over Internet Protocol (IP) See *VOIP*.

VOIP Voice Over Internet Protocol (IP). The ability to carry voice signals on packet-switched networks based on the Internet Protocol.

W3C® An acronym for World Wide Web Consortium. A body that develops specifications and technologies related to the World Wide Web.

WAAS An acronym for Wide Area Augmentation System. A system designed to boost the accuracy of GPS systems by transmitting GPS correction signals from a set of geosynchronous satellites. At the time this book was written, WAAS was still under development, but was available for public use.

WAP An acronym for Wireless Application Protocol. An open, global specification that defines one way in which online services can communicate with wireless devices.

War Driving Driving through an area with equipment set up to detect and connect to wireless networks.

WCDMA An acronym for Wideband CDMA. A 3G standard that offers a data rate of up to 2 Mbps, while making highly efficient use of the RF spectrum. GSM systems will migrate to WCDMA technology when they want to provide 3G service. The standard is fully compliant with IMT-2000, and forms the air interface technology for UMTS and ARIB. WCDMA systems are under test at various locations around the world. Also written *W-CDMA* or *wCDMA*.

WDF An acronym for Wireless Data Forum. A leading wireless data and mobile computing trade association.

WECA An acronym for Wireless Ethernet Compatibility Alliance. Among its other functions, this organization certifies 802.11b devices as Wi-Fi-compliant.

WEP An acronym for Wired Equivalent Privacy. An optional security technology defined in the 802.11 specification. When implemented, WEP provides wireless security equivalent to that of a physical cable.

WID An acronym for Wireless Information Device. A term used to describe wireless devices that combine both voice and data capabilities in one small, portable device. WIDs are now commonly called Smartphones.

Wide Area Augmentation System See *WAAS*.

Wi-Fi An acronym for Wireless Fidelity. A trade name for devices that meet the IEEE 802.11b standard and are interoperable. Wi-Fi compliance is certified by WECA.

Wideband CDMA See *WCDMA*.

Wired Equivalent Privacy See *WEP*.

Wireless Application Protocol See *WAP*.

Wireless Data Forum See *WDF*.

Wireless Ethernet Compatibility Alliance See *WECA*.

Wireless Fidelity See *Wi-Fi.*

Wireless Information Device See *WID.*

Wireless IP An acronym for Wireless Internet Protocol. Another term for *CDPD*.

Wireless Local Area Network See *WLAN.*

Wireless Markup Language See *WML.*

Wireless Service Provider See *WSP.*

Wireless Transport Layer Security See *WTLS.*

WLAN An acronym for Wireless Local Area Network. A network connected together by short-range wireless links such as Wi-Fi. Sometimes also written *W-LAN.*

WML An acronym for Wireless Markup Language. A markup language based on XML that is used to create format information for use with WAP-capable devices.

World Wide Web Consortium See *W3C.*

WSP An acronym for Wireless Service Provider. A company that provides wireless connectivity to users of wireless devices. Analogous to an Internet Service Provider (ISP).

WTLS An acronym for Wireless Transport Layer Security. The security model used by WAP. WTLS uses less resource-intensive encryption algorithms than approaches like SSL.

XML An acronym for eXtensible Markup Language. XML was created to allow the use of richly structured documents over the Web. Several markup languages of particular interest to wireless device users are based on XML. Examples include SyncML and WML. The XML specification was created by W3C.

Index

Numerics

C

continued

continued

continued

H

I

continued

N

O

P

continued

Q

R

S

X

Y

Z